Valuing Animals

Animals, History, Culture

Harriet Ritvo, SERIES EDITOR

Valuing Animals

Veterinarians and Their Patients

in Modern America

SUSAN D. JONES

THE JOHNS HOPKINS UNIVERSITY PRESS
Baltimore & London

© 2003 Susan D. Jones
All rights reserved. Published 2003
Printed in the United States of America on acid-free paper
9 8 7 6 5 4 3 2 1

The Johns Hopkins University Press
2715 North Charles Street
Baltimore, Maryland 21218-4363
www.press.jhu.edu

Library of Congress Cataloging-in-Publication Data

Jones, Susan D., 1964–
    Valuing animals : veterinarians and their patients in modern America /
Susan D. Jones
        p. cm. — (Animals, history, culture)
Includes bibliographical references (p.   ) and index.
    ISBN 0-8018-7129-8 (hardcover : alk. paper)
    1. Veterinary medicine—United States—History. 2. Domestic animals—Social
aspects—United States—History. 3. Human-animal relationships—United States—
History. 4. Animal welfare—United States—History. I. Title. II. Series.
    SF623 .J65 2002
    636.089′0973—dc21
                                2002003253

A catalog record for this book is available from the British Library.

For Kevin

Contents

Preface   ix

Introduction   1

CHAPTER ONE
*Doctoring a Nation of Animals at the
Century's Turn*   10

CHAPTER TWO
Valuable Patients
Horses and the Domestic Animal Economy   35

CHAPTER THREE
The Value of Animal Health for Human Health   63

CHAPTER FOUR
The Value in Numbers
Creating "Factory Farms" at Midcentury   91

CHAPTER FIVE
Pricing the Priceless Pet   115

CHAPTER SIX
Reconciling Use and Humanitarianism   141

Notes   155
Essay on Sources   201
Index   209

Preface

Growing up on the family farm in the early 1970s, my brothers and I inhabited a world both magical and brutally practical. We spent hot afternoons digging crinoid fossils, arrowheads, and clay pipe stems out of the cool bed of a creek, probing down through cow tracks filled with water in the rocky sand. Evenings brought glorious sunsets visible for miles over gently rolling fields and the clean smell of tall grass in the waterways. We also sweated and prickled as we worked in the hayfield, choked on the dust raised by the tractor in the early summer, and cried as we killed chickens or shoveled up the remains of calves hit by cars on the road. We raised dogs and cats and attempted to tame Angus yearlings; the greatest day of my young life was spent bringing home my very own horse.

We lived intimately and (for the most part) happily with the people, animals, and plants that made up our world. Yet the type of experience we had was rare in our time and might better have characterized the lives of American children 70 years before. Like us on the farm, those earlier children would have been surrounded by animals every day, even if they lived in cities. They would have walked and traveled in the company of horses, encountered wandering dogs and cats, and probably saw chickens or a milk cow in at least one neighborhood backyard. Young people were not spared the realities of human dependence on animals—even urban children had probably observed the milking of a cow, the butchering of chickens, and drivers whipping their lathered horses. Intimacy, interdependence, and visibility characterized their relationships with the animals that surrounded them.

A fast forward to the beginning of the twenty-first century shows my brothers and me, now adults, living in midsized to large cities, along with most of the American population. We rely on internal combustion engines (and cheap gasoline) for transportation. We purchase our food already killed, cleaned, and shrink-wrapped in the grocery store, and our collective livestock amounts to three cats. Our daily interactions with animals have narrowed since we left the farm; our interdependence with them has become submerged. I have often reflected on my own family's experience as a sort of time-squeezed microcosm of Americans' shifting relationships with domesticated animals over the past century. Tenuous as this connection may seem, the change in these relationships vividly reflects basic alterations in the daily experience of American life. Uncomfortable with the implications of our dependence on some animals, we have banished them from most of our daily experience. At the same time, the qualities of other animals have helped them to become more visible participants in American culture and society of the late twentieth century.

In this book I explore the changing relations between Americans and their domestic animals throughout the past century to help us understand the implications of transformations from backyard milk cows to the cartons in the supermarket, and from lathered, whipped horses to the dog wearing a red plaid coat on his winter morning walk. The book examines contradictions in our treatment of domesticated animals. How have different groups of Americans valued various groups of animals? We cannot assume that cows have fared the same as horses, or dogs, or chickens any more than we can avoid studying how rural versus urban people, women and men, and different ethnic groups and professions have configured their relationships with animals over time. Moreover, I treat animals as more than just symbols within human culture. Their unique situation as living beings who are considered and used as property urges us to explore both the ideas and the practices that have shaped their lives. The position of animals in human society and culture has spawned tremendous efforts to make them into social and cultural commodities through complex and often not very visible processes that this book brings to light.

Many of these processes have been directed over the past century by experts in the care and use of animals. I look specifically at the role of veterinary medicine in the changing relations between Americans and their domestic animals. As the following chapters show, veterinary scientists and practitioners of veterinary medicine provided many of the ideas and tools that reshaped the lives and roles of animals; they were a primary force in validating and sometimes establishing the nuances of animal value. In turn, the value placed by Americans on animals restructured the veterinary profession. During a century in which applied science acquired tremendous power and authority, veterinarians found that animal value could both distinguish their work from that of other animal scientists and limit what they could accomplish. Certainly, veterinarians' activities reveal the positions of various animals in society and vice versa. Without studying both, we stand to lose our perspective on either—a grand historical blunder, considering the importance of animals in the construction of American infrastructure and identity.

Despite the importance of these issues, most recent historians have paid little attention to animals. There are notable exceptions, some of whom have contributed directly to the genesis of this book. For patiently reading draft chapters and helping me to clarify my argument, I thank John P. Jackson Jr., J. F. Smithcors, Elizabeth Hanson, and particularly Virginia DeJohn Anderson, whose own work in the history of human-animal relationships and marvelous sense of humor have nourished me at the University of Colorado. Editor Robert J. Brugger and series editor Harriet Ritvo guided me and enthusiastically supported publication of this book. The suggestions of an anonymous reviewer and copyeditor Ruth Haas improved the manuscript. Many colleagues have been willing to discuss "animal ideas" with me: my fellow graduate students Chris Feudtner, Mark Hamin, Jennifer Gunn, Carla Keirns, Jean Silver-Isenstadt, Susan Miller, Elizabeth Toon, and Doug Tarnapol at the University of Pennsylvania; historians Larry Carbone, Elizabeth Hanson, Rob Kohler, Susan Lederer, John Parascandola, and Barbara Rosenkrantz and anthropologist Elizabeth Atwood Lawrence (all of whom have written about animals); and veterinary historians Robert Dunlop, Peter Koolmees, Phyllis Hickney Larson, and Leo Lemonds.

Many of the ideas in this book have been presented at conferences and seminars, and I thank those who have questioned and commented on them. I am pleased to acknowledge my intellectual debt to Charles Rosenberg, whose generous heart and far-ranging mind guided my doctoral dissertation and much work since then. I have benefited tremendously from Phil Teigen's keen interest in animal history, kind advice, and willingness to serve on conference programs with me. Of course, given the excellent advice of these scholars, this study should be unblemished. I am sure that I echo the plaints of many first-time authors when I lament the fact that I lacked the time to respond to and incorporate all the suggestions that I received. Thus, whatever is inconsistent, awkward, or incorrect is my responsibility alone.

This book would not have the shape it does without the work of scholars in closely related fields. James Serpell introduced me to the sociologically oriented study of human-animal relationships. By assigning Viviana Zelizer's book, *Pricing the Priceless Child,* in his U.S. social history course, Michael Katz first exposed me to Georg Simmel's theories of economic sociology, which anchor my analysis of animal value. William Cronon has inspired so many scholars interested in the connections between humans and what they see as "nature"; all of us who write about animals owe a debt to him and to other environmental historians who have insisted that the nonhuman components of our world are crucial to historical study.

I also am grateful for the support of a number of institutions and the people who run them. The University of Pennsylvania and the University of Colorado provided grants, resources, and teaching leave. The expertise of archivists and librarians made my research possible. I thank Tyler Walters and the staff at Iowa State University; Joe Schwartz at the National Archives (College Park, Maryland); Jane Thompson and Lynette Westerlund at the University of Colorado; the staff at Colorado State University, the University of Pennsylvania's Fairman Rogers Collection, Western Michigan University, and Cornell University's Kroch Library; and Susanne Whitaker and her staff at the Flower-Sprecher Veterinary Library, Cornell University, who allowed me to use rare unprocessed materials from the library's archives. The American Anti-Vivisection Society kindly allowed me access to their files and journals. I found further important material in the City Archives of Philadelphia and the Denver Public Library.

Finally, I thank my brothers, Ed and Jim Jones, for those memorable afternoons down at the creek, and my parents, Phil and Judy Jones, for raising us on the farm and simultaneously preparing us for the wider world. Words are inadequate to express my gratitude to Kevin Reitz, who has supported me in every way. An early convert to the sociocultural history of human-animal interactions, he has read much of this study and contributed intellectually to it. This has been our project almost as much as it has been mine.

Valuing Animals

Introduction

Americans live closely with, and depend on, domestic animals. We eat and wear the products of their bodies and share our homes and personal spaces with them. Yet few of us have thought much about the inconsistencies inherent in our relationships with domestic animals. The creatures populating the domestic animal economy have contributed greatly to our social landscape and permeated our cultural identity. At the same time, most Americans have little contact with domestic animals aside from pets; we do not spend time with farm animals or observe the processes required to turn them into food. Why do we eat some animals and treat others as members of our families? How have domestic animals come to occupy their present places in American culture and society?

To deepen our understanding of modern American relationships with domestic animals, this book explores the history of how Americans have assigned "value" to horses, cattle, dogs, cats, and other animals. The term *value* has had very different meanings in different intellectual contexts. Nineteenth- and twentieth-century political economists, including Karl Marx and Georg Simmel, probed the social relations inherent in the value of everything from labor to the purchase of objects (and even the value of humans). Philosophers and historians (Friedrich Nietzsche most prominently) insisted that value also encoded the ideals and motivations of individuals and the telos of a whole culture. Combining these two characterizations of value suggests a fruitful way of studying human-animal relationships. It helps us to explain Americans' seemingly inconsistent treatment of animals over time. It also focuses on the attributes of the animals themselves—their health, behavior, strength, speed, rarity, loyalty, and so on—as factors in their perceived worth.

Over the past century, Americans have learned to value the attributes of animals in part by listening to the advice of experts. Experts have been an important source of answers to questions about how animals should be used in American society and culture: What foods of animal origin are appropriate to eat? What definitions and concerns for bodily health and well-being have been attached to economically productive animals? What place should companion animals have in families and communities? Overall, no experts have been so influential in these domains as veterinary scientists and veterinarians. A little over a century ago, the veterinary sciences became an organized field and veterinarians formed a profession. These scientists have presented themselves ever since as mediators of human-animal relationships. By studying the "very nature" of domestic animals' bodies and behaviors, veterinary scientists have claimed a position of primacy in judging how these animals should be valued and used.[1]

A closer look at their activities shows that the reverse has also occurred: changes in animals' social and cultural value have influenced the sociological and intellectual development of the veterinary sciences and veterinary medicine in the twentieth century. Until the 1890s, veterinary medicine boasted few of the characteristics of a modern profession. By 1910, veterinarians had successfully campaigned for state licensure and regulation of their professional interests; they had established successful schools and training programs and gained control of a branch of the U.S. Department of Agriculture (USDA). They sought to accomplish all of these professional goals by arguing repeatedly that veterinary medicine protected animals' economic worth and other valuable contributions to American life. At the beginning of the twentieth century, veterinarians focused on animals that provided transportation, food, and other material resources. With time and changing circumstances, however, veterinarians also realized the importance of pets—animals that were valued for companionship and as symbols of humanitarianism. By the end of the century, veterinary medicine and the veterinary sciences concentrated largely on the health of companion, or pet, animals.[2]

The professional self-identification and intellectual program of veterinarians altered tremendously over time to reflect Americans' cultural (as well as material) interests in animals. A range of cultural values, along with practical and economic values, influenced people's relationships with their domestic creatures, and all were contained in what is termed here the "domestic animal economy." The term *œconomy* originally meant "a set of household relations," and this concept has been enlarged by scholars to include the economic and social relationships of much larger groups of people (or other organisms). Historian E. P. Thompson defined the customs and expectations of English working-class bread rioters as a "moral economy"; Donald Worster has even included all the components of the natural world in his study of ecology, or "nature's economy."[3] The domestic animal economy can be defined by its three components: (1) humans and animals tied together by a market or financial relation, (2) cultural beliefs about animals, and (3) the customs of human-animal interactions in a particular time and place. All of these components contributed to Americans' changing relationships with domestic creatures.

Veterinarians have sought to address Americans' uncertainty about the "proper" human-animal relationship as the ideological driving force of their profession. They did not pretend to be philosophers, but operated as rationalists meeting social needs. Over the past century animals were considered nonhuman but were often anthropomorphized; some animals were eaten while others were treated as family members. Veterinarians provided important tools that allowed these inconsistent human-animal relationships to coexist. They validated the use of animals as economic commodities even while praising themselves and most of their fellow citizens for their compassionate attitudes toward domesticated creatures. I argue that veterinarians' contributions to the reconciliation of animal use with concerns about morality shaped the development of large-scale production of animals for food, commercialized pet keeping, and other features characteristic of Americans' relationships with domestic creatures in the twentieth century.

Why focus on human-animal relationships in the twentieth century? Animals played a more visible role in American life in the nation's earlier years, when pigs still roamed the streets, "horse power" was a literal term, and most people lived in rural areas. This was not so in the twentieth century, which has been characterized as the triumph of urban bureaucratic control, the progress of technology replacing "archaic" practices such as horse-powered transportation, and the distancing of Americans from a slower, dirtier, more parochial time. Yet there is no more interesting time to study human-animal relationships in this country because they changed so greatly during "the modern century," both in terms of social practice and in how Americans have shaped individual and collective ideas about health, humanitarianism, and prosperity associated with animals.[4]

Social scientists, arriving at this realization, have begun in the past 15 years to explore the relationships between humans and animals. They have sought to explain people's attitudes toward animals, the ethical dilemmas posed by human use of animals, the cultural spaces occupied by animals, and the parameters of human companionship with animals. Likewise, a handful of historians have examined these relationships in studies of specific times and places, such as seventeenth-to-nineteenth-century Britain and France; or in historical examinations of a specific problem, such as animal welfare or rights. Historians have evinced some interest in animals because of their contributions to twentieth-century agriculture and science in the United States.[5] While exposing some of the difficulties of studying animals and their place in modern society and culture, all of this work has influenced *Valuing Animals*.

This book addresses a new area of inquiry: the contextual history of comparative medicine. "Comparative medicine," referring to veterinary and human medicine, is certainly not a term of my invention; it dates back at least to the early nineteenth century. The use of "contextual" and "comparative" to describe histories of animal health and disease sends some important signals. First, these terms draw historians of veterinary medicine away from the celebratory narratives of scientific progress that have characterized most of the literature. Second, they reflect a reliance on the methodologies of social and cultural history and recent work in the history of medicine. Finally (and this book is insistent on this point), these terms remind us that it is impossible to understand the development of veterinary medicine, and the changing sociocultural roles of animals, as processes isolated from each other.

Several questions drive this study of human-animal relationships and veterinary medicine. The first of these gets at the heart of the historical enterprise and at who (or what) counts as a "historical actor." For historians, writing a history about animals as well as about their relationships with people presents certain difficulties. Animals are perpetual "others," doomed to have their interests represented to humans by other humans. Thus writing history from their point of view would seem to be an impossible task. The best that historians can do is to make approximations. These approximations usually involve biological and ethological knowledge—the scientific frame we place around our observations of animals' bodies, minds, and behaviors. Environmental historians are familiar with using these types of knowledge, and this historical subdiscipline provides a foundation on which historians can and must build. As Elliott West has suggested, scholars should try to consider the perspectives of both human and nonhuman actors when constructing historical arguments. It does not necessarily follow that such arguments are teleological or anthropomorphic; certainly this will depend upon how they are made. Presenting animals as historical actors may not appeal to many historians, but it is nonetheless necessary when considering animals or the relationships between humans and animals historically.[6]

A second important question, and one obvious from the title of this book, consists of how historians can measure the ways that historical actors value things. In this study, the work of economic sociologists, particularly Georg Simmel, provides the model for applying value to animals. Writing at the end of the nineteenth century, Simmel declared that "economic forms themselves are recognized as the result of more profound valuations." He insisted that the economic form of value—price—disguised a complex set of psychological desires, cultural demands, and social constraints peculiar to a time, a place, and even an individual. A product of his time, Simmel was especially interested in psychology and metaphysics, yet his analysis did not neglect the more practical attributes that might make a thing more desirable to a potential purchaser. The price of anything, then, had to be analyzed by exploring the values attached to it. Simmel went on to dissect other intricacies of the meaning of value for price, but for present purposes, his general theoretical line of inquiry proves very helpful in understanding the sociocultural role of animals. Other scholars have used Simmel's ideas in historical studies; for example, Viviana Zelizer has applied them to the changing value of children in American society.[7]

When considering animals, the case for understanding how sociocultural values are reflected in price is particularly urgent. During the colonial days of English common law, and codified since by the laws of the United States, animals were legally categorized as chattel, or property.[8] Because of this characterization, money has been perhaps the most common mediator of Americans' relationships with their domestic animals. Thus, the third important question that this study asks is how the cultural beliefs and social needs of different groups of Americans have been attached to different groups of domestic animals over time. Understanding this aspect of animals' economic value is the only way that characteristics such as affectionate disposition, health, strength, or pedigree can become part of their official (economic) worth and identity. It also serves as a reminder that different groups of (and even individual) animals could be valued in different ways. Thus, national figures on the average price of a horse or cow represent only a starting point for understanding its value, because animals had (and have) surprisingly complex identities as individuals and groups. It is tempting to reduce the value applied to animals to a monetary price; indeed, animals have always been subject to such appraisals because of their status as chattel. Even so, economic reductionism in this case provides little explanatory power for historical analysis. People often interacted with animals in ways that were inconsistent with the dictates of property or profit maximization, and historians must try to understand why.

This realization raises a fourth important question: who determined value? Certainly every transaction meant that the humans involved arrived at individual judgments of an animal's value. Yet national and regional currents of social need, cultural belief, and expert advice, interpreted locally, provided a structure and suggested methodologies for judging animal value. I argue specifically that the veterinary sciences and veterinary medicine functioned as major expert advisors on valuing animals. This is not a history of veterinary medicine in the traditional sense; it does not strive for the inclusion of all important developments and people. The greater story lies in how veterinarians contributed to and manipulated animal value in order to claim a place as indispensable mediators of human-animal relationships. The veterinary sciences and veterinary medicine sought to apply a scientific intellectual framework to understanding the bodies, behaviors, abilities, and commercial uses of animals. Insofar as this type of information was used by agriculturalists, food production businesses, and others, it was often veterinarians who transmitted it. The agricultural colleges, the federal government's meat inspection service, the horse-racing industry, pet owners—all depended on veterinarians to teach, judge, and medicate.

They operated within an increasingly bureaucratic and "scientized" society: a society that has been called "modern," according to Max Weber's definition. Writing in the years around World War I, Weber identified an assemblage of social institutions and processes that together were creating a new, "rational" (or rationalized) society. Conformity, control, and scientific and technological production would solve social problems and order social interactions, including those between humans and animals. While this book is not explicitly concerned with defining "modernity," Weber's ideas serve as a useful framework in which to understand how the relatively small profession of veterinary researchers and practicing veterinarians could have had such a large impact on the lives of Americans and their domesticated animals. Through the reinforcing effects of scientific authority and state sanction, the development of veterinary medicine heavily influenced the domestic animal economy in the United States.[9]

As Weber might have predicted, in the first three decades of the twentieth century veterinary leaders used the power of the state as a tool to regulate their profession, enforce their role as animal experts, and capture scientific legitimacy. The federal Bureau of Animal Industry (BAI) and the veterinarians running it proved to be powerful influences in all three areas from the BAI's founding in 1884 to the depression years. It was the BAI, not the American Veterinary Medical Association (AVMA) or the loose consortium of veterinary college faculty, that most heavily regulated veterinary education from the turn of the century through the 1920s. The BAI spearheaded legislation that put veterinarians in control of the biologicals used to test and treat animal diseases around the nation. Finally, BAI laboratories were the single most important site of veterinary research, some of which revolutionized the international scientific understanding of disease mechanisms. All such organizational accomplishments meant that individual veterinarians, despite their reputation for stubborn independence, learned and practiced at least in part according to this federal bureau's dictates.

However, a more powerful influence on the veterinary profession existed: the nation's need, and individuals' willingness to pay for, veterinary services. The BAI argued every year in its annual report that the value of the nation's animals merited increased appropriations for the bureau (making the assumption, of course, that more veterinarians and more veterinary knowledge *would* save the lives of more animals and fill the pocketbooks of their owners). Not surprisingly, the economic value of animals' lives also structured the practices of individual veterinarians. All the serums and medications in the world were useless if their expense outstripped the value of the animal and the owner declined to use them. Owners' valuation of their animals, veterinarians knew, represented the most serious limitation to their ability to apply scientific knowledge to animal health and disease.

It also represented veterinarians' greatest opportunity to shape Americans' relationships with domestic creatures, and this point became particularly clear with the events that followed World War II. In 1953, the BAI was dissolved and its administrative and research functions were divided up among several federal agencies. Disease eradication programs championed by the BAI ended or were restructured, and locally practicing veterinarians lost their jobs as state functionaries. Those who had concentrated on food-producing animals in their practices not only suffered the loss of federal contracts, but also observed a decline in the population of small livestock producers who had been their primary clients. Canny practitioners, however, had already noticed the development of a new business opportunity: translating the sentimental value that Americans attached to companion animals, or pets, into a mandate for medical care. Despite a traditional ethos in veterinary medicine that eschewed valuing an animal emotionally, professional leaders and practitioners worked in the mid-twentieth century to support and shape Americans' interest in pets. Veterinary researchers consciously redefined their scientific enterprise in order to differentiate sentimentally valued pets from laboratory animals (which were often of the same species).

Thus, animal value, rather than scientific or professional advancement, most closely shaped the development of American veterinary medicine in the 1900s. In return, the veterinary sciences and veterinary medicine serve, in the words of Charles Rosenberg, as "multidimensional sampling devices" for understanding the changing relations between Americans and domesticated creatures.[10] Neither can be truly understood outside the context of the other in the twentieth century. The following chapters provide selected studies that illustrate these points and help us to answer the underlying questions about Americans' relationships with domestic creatures. Each chapter is thematic and explores a particular group of animals and its role in American society. While some of the chapters overlap in time period, overall the narrative moves forward from about 1900 through the 1990s. It highlights crucial junctures at which transformations in animal valuation and the development of the veterinary sciences and veterinary medicine influenced each other: the transition from horse power to motorized vehicles, public health concerns over animal food products, the rise of "factory farms," and the increasing importance of pet practice to veterinary medicine.

As this book illustrates, veterinarians learned to recognize and shape types of animal value and to transform their professional focus accordingly. Rather than relying solely on scientific developments, the intellectual and sociological successes of veterinary medicine have also depended upon its ability to mediate Americans' concerns about their relationships with domesticated creatures. Indeed, veterinarians have worked over the past century to reconcile the exploitation of animals for food, work, and companionship with Americans' need to feel morally comfortable with those uses. The balance between these two conflicting needs has been renegotiated several times, and will continue to be negotiated in the future. Animal value is dramatically changing again at the beginning of the twenty-first century. To understand what this is likely to mean, we must go back in time, to the streets of a nation crowded with animals and economic opportunities for their caretakers.

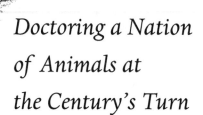

CHAPTER ONE

Doctoring a Nation
of Animals at
the Century's Turn

The year is 1895, early morning on a dusty street in an American town. Besides the dust, the smell of warming horse manure assails the nostrils. The morning sounds include the clucking of chickens in most backyards, the clip-clop of horses' feet, and dogs barking as they trot boisterously down the street. The rising sun is reflected in the coppery and dusky coats of the horses and reveals cats slinking behind barrels and under porches. Like squabbling sisters, two horses jostle raucously at the tie-rail next to the board sidewalk. With few people on the street at this early hour, the animals have claimed it as their own.

A century later, a time-traveler from this 1895 scene would scarcely recognize this town for obvious reasons: the smell of auto exhaust rather than horse manure, the hard concrete sidewalk, neon signs, and the virtual disappearance of animals. Even in 1895, however, our visitor could have observed the stirrings of change that would connect the two street scenes. In retrospect, historians have characterized the early twentieth century as (among other things) the time in which Americans placed their faith in social improvement through technology, professional expertise, and managerial bureaucracy; reevaluated the role of government regulation in their lives; and struggled to live within an increasingly pluralistic population engaged in creating a mass popular culture. Relationships between people and their domesticated animals were transformed accordingly. Internal combustion technology replaced horse transportation, to the relief of many and the regret of some. Animals used for food—cattle, hogs, sheep, and others—retreated completely from most Americans' thoughts and experiences. Although they had deeper roots, these transformations and others within the domestic animal economy effectively altered Americans' lives in the twentieth century.

This chapter describes a turn-of-the-century world teeming with animals that would have been familiar to Americans of the time, no matter where they lived. It also traces the intimate connections between these animals' changing roles in American society and the simultaneous development of the veterinary sciences and veterinary medicine. The needs of animal populations and their owners stimulated ambitious veterinarians to work at forming a profession. Veterinarians positioned themselves to care for the most economically valuable animals by adopting particular scientific ideas and practices that promised to safeguard the animals' health and their owners' profits.

The Transformation of Horse Doctoring

Throughout most of the nineteenth century, many individuals who doctored animals professionally obtained their training through experience or apprenticeship, as did physicians and lawyers. Hanging one's shingle out as a "veterinarian" required no legal sanction, merely an ability to persuade potential customers that one's services were worth the price. Some veterinarians had received training in a formal veterinary school or program; the course was often brief, however, and the quality of education varied tremendously.[1] By the end of the 1890s, Americans could attend veterinary schools in Toronto, Montreal, New York City, Chicago, Detroit, Kansas City, Indianapolis, and Grand Rapids. Universities such as Harvard, Iowa State, Cornell, Ohio State, Washington State, and the University of Pennsylvania had veterinary departments or colleges. Nevertheless, most of those doctoring animals came to their vocation through experience in animal-related occupations such as driving, farriery (horse podiatry), or animal breeding; or as the result of working for a practicing veterinarian.[2]

The culture of animal doctoring borrowed heavily from that of the stable and barnyard, and the usual denizens of both occupied lowly positions on the social ladder. The duplicitous "cigar-chomping horse doctor" and slovenly "cow leech," both stereotypes of animal healers, derived from the images of drunken grooms and vulgar farmhands.[3] R. J. Dinsmore, who worked in his father's livery stable as a boy and later graduated from Harvard Veterinary School, remembered the early 1890s' image of the "horse doctor" this way: "Nobody was laughed at more than the horse doctor. Horse doctors were supposed to be a coarse, ignorant group who had made a failure of blacksmithing or farming and had turned to 'doctoring.' That they actually knew anything about medicine was an absurd notion. . . . Most were not real veterinarians, but farriers."[4] Dinsmore's characterization of animal doctors pointed to not only their lack of education but also the rough-and-tumble world from which they came. The same brush of low social status tarred competent and incompetent practitioners alike.[5]

Before the 1890s, such typecasting often foiled ambitious veterinarians desirous of establishing schools, journals, and other components of professionalization. Writing in 1883, Robert Jennings remembered that his efforts to establish the first veterinary school in Philadelphia foundered because "young men of education and respectability would not engage in a profession of so low a standing."[6] Also in 1883, congressional sponsors of the Bureau of Animal Industry charter found themselves defending the qualifications and respectability of "graduate veterinarians" as the House of Representatives debated the necessary appropriations. Representative William H. Hatch felt obliged to answer his fellow legislators' "sneers and jeers" aimed at "horse doctors."[7] Created only after a heated discussion, the BAI's charter stipulated that its chief officer had to be a "competent veterinary surgeon," thus ensuring veterinary (rather than medical or agricultural) control of investigations into animal diseases and of government jobs. Veterinarians and their supporters won the right to administer the BAI only after successfully contravening some House members' characterization of them as "illiterate horse doctors."[8]

Animal doctors, by and large, had also adopted a particular ideology and practice based on gendered assumptions about their work.[9] The culture of the stable and barnyard was a very masculine one, and with the exception of certain limited tasks, it generally excluded women.[10] As historians have shown, women who were determined to become physicians successfully argued that their special qualities as nurturers would make them good doctors, especially for other women and children.[11] Women interested in practicing veterinary medicine, however, had difficulty in finding a similar argument. In 1897, the *American Veterinary Review* editors, while being careful not to state an outright prohibition against women veterinarians, asserted that "veterinary surgery is of all the learned professions the one least adapted for women." Reprinting excerpts from a letter written by a British veterinarian, the American editors supported the point of view that women should "fulfill those duties for which they are fitted by nature." Veterinary medicine, born of masculine barnyard culture, most emphatically did not allow for the expression of the feminine nature.[12] The idea of women veterinarians not only violated masculine livestock culture, it also threatened the professional aspirations of veterinarians. Women, widely assumed to be unsuitable for serious study or work outside the home, had to be excluded from a discipline that was attempting to become a selective, learned profession.[13] Furthermore, a significant portion of veterinarians' bread and butter came from a surgical procedure that could only be seen as objectionable for women to perform—the castration of male animals. Male veterinarians argued that women did not belong in the rough world of the barnyard and threatened any woman daring to trespass with the loss of her "delicacy of feeling"—her femininity.[14]

The British writer and American editors also linked femininity, women animal doctors, and animals valued emotionally rather than for economic worth. In doing so, the editorial exposed veterinary culture's underlying unsentimental attitude toward its patients: "If the practice of veterinary surgery consisted in making a round of visits among lap-dogs, . . . and simply diagnosing their diseases . . . then, and only then, the profession might be a suitable one for women possessed of any delicacy of feeling."[15] Professional animal doctors seldom, if ever, cared professionally for pets at this time and disdained these animals' social location in "feminine drawing rooms." Condemning the association of his occupation with such frivolity, one American veterinarian warned that if pets became patients "we may soon expect to see the advertisements of our veterinary colleges in the ladies' magazines."[16] Since most patients were animals valued for their productive abilities as workers or food, an unsentimental professional ethos made sense on a number of levels. It created a comradeship with most livestock owners, who were interested in protecting their valuable animal investments, and thus fulfilled the profit motive of barnyard culture. Veterinarians could also easily make an uncomplicated economic argument for the worth of their services to individual owners and the livestock industry as a whole. Finally, the profession supported the unfettered use of horses as workers and cattle, hogs, sheep, and poultry as food producers. These animals had economic and practical value, and veterinarians based their strategy for professional success on this actuality.

This stance of veterinary culture encouraged a reliance on a hard-edged economic analysis and what we now view as unpleasant or even brutal treatment practices. Similar to the "heroic" treatments of nineteenth-century physicians and surgeons, the procedures of animal doctors often consisted of burning or firing, blistering, surgery without anesthesia, bleeding, and strong purgatives. These treatments reflected tradition and arose from certain theories about the causes of disease, especially the idea that an imbalance of bodily humors caused pain and disease. They also created dramatic effects that could be used to persuade animal owners of the healer's ability to intervene.[17] But perhaps just as important to veterinarians, such treatments reinforced the essential unsentimentality of their work.

I do not mean to imply that animal doctors were sadistic or deliberately barbaric; rather, heroic treatments served tradition, clients' expectations, and self-image simultaneously. Three examples of such treatments—burning, firing, and blistering—involved heating special iron implements and applying them to the animal's affected part (often a lame leg in horses). Neurotomy, the established treatment for chronic or intractable lameness, involved severing the nervous supply to the affected hoof or lower leg so that the animal could not feel the pain of its injury. Castration and other surgical operations required the use of ropes and strong assistants to "throw" or "cast" the animal, forcibly immobilizing it while the surgeon did his work (not surprisingly, speed and dexterity were much prized).[18] These procedures all belonged to the accepted standard repertoire of veterinarians' practice, and all illustrated veterinary culture's fundamental hostility to coddling its patients.

Altering this view would be a project of veterinary leaders decades later, but in the 1890s their efforts at professionalization focused on defining and protecting their roles as doctors to economically valuable animals. The marketplace for animal care was a crowded one, with land-grant colleges turning out educated farmers and livestock husbandmen. Aspiring veterinarians needed special knowledge, skills, and professional authority. The "horse doctor" stereotype and the persistence of heroic treatments meant that the transformation of animal doctoring into the twentieth-century veterinary profession would proceed slowly. As with other disciplines interested in becoming professions in the last quarter of the nineteenth century, the first step involved founding veterinary schools to train and enculturate new members of the discipline.[19] The loss of many valuable horses to glanders during the Civil War and the periodic outbreaks of epizootics among cattle contributed to the arguments that aspiring veterinary educators used to justify the establishment of veterinary schools. In 1865, veterinarian A. S. Copeman gave the inaugural lecture at the opening ceremonies for the New York College of Veterinary Surgeons. He emphasized that "the great problem of veterinary medicine is how to preserve the health of domestic animals and thereby increase the wealth of the nation."[20] From their earliest inception, veterinary schools in America existed to serve their profession's interest in caring for the most economically valuable animals in the domestic animal economy.

Although they were a relatively new phenomenon, veterinary schools exerted an increasingly powerful influence on professionalization as states began to regulate the health-care market for animals in the 1890s. Ambitious young veterinary school graduates and their mentors drafted and lobbied for state legislatures to pass laws restricting the definition of "veterinarian," and the right to work as an animal doctor, to registered and licensed practitioners. Anyone who had refused to register at the time a state adopted such an act, and any subsequent applicants who had not graduated from a veterinary school, could no longer legally practice what was now called "veterinary medicine." States containing veterinary schools or large numbers of graduate veterinarians passed the first restrictive legislation.[21] By the beginning of the twentieth century, these legal restrictions ensured that a majority of licensed veterinary practitioners would be graduates of one of the several veterinary schools. Once protected by state practice acts, veterinarians felt that they could make a legitimate claim to proprietary knowledge and competence that justified their role as guardians of the health of valuable animals.[22]

Given this understanding, registered practitioners of veterinary medicine debated the ideological goals and exact parameters of their professional role in the 1880s and 1890s. They had divergent interests based on their training, experience, and practical needs. Most of the rank and file concentrated on treating the medical problems of horses, and the daily grind of attracting and retaining clients remained paramount. This routine was occasionally punctuated by issues of food and environmental safety that threatened human health and animal epidemics that threatened economic ruin. Their veterinary school professors and other professional leaders provided guidance and advice through local meetings and national journals (such as the widely read *American Veterinary Review*). These leaders had a disproportionate influence on professional strategies, but they could not ignore the needs of the practitioners. By the end of the 1890s, veterinary schools' leaders had begun to characterize the profession in terms of economic stability for practitioners and a responsibility to the improvement of American society at large.[23]

Leaders of the profession had various visions of social responsibility, however, and they aired their differences openly. For example, veterinarians struggled with reconciling their allegiance to the agricultural livestock economy (and its profits) and opportunities to be identified more closely with physicians and public health officers. Although veterinarians understood the importance of maintaining the health of valuable animals, many leaders cast their profession as guardians of human health as well. At least one veterinary school won state appropriations based on concerns about diseases, such as tuberculosis, that were presumed to be transmitted between humans and animals. But the needs of public health and profit could clash if animals used for food or work endangered human health. Professional leaders had to navigate between both visions of social responsibility—the health *and* wealth of the nation.[24]

The establishment of the BAI pitted the mission of veterinary medicine as a public health agency against its interest in protecting the profits of the livestock economy. Daniel E. Salmon, the Cornell-educated veterinarian who was appointed the BAI's first chief in 1884, was among those who felt that the bureau belonged under the purview of the U.S. Department of Agriculture. Salmon emphasized that the main concern of veterinarians, the BAI's largest group of professional employees, lay with maintaining the health of economically valuable livestock (his argument also reflected the importance of livestock interests in lobbying for an increase in BAI appropriations). Other veterinary leaders, including French-trained Alexandre Liautard, opposed Salmon's view. Liautard wanted the BAI's work to be supervised by the National Board of Health then in existence; he wrote that "the investigation of animal diseases is a matter entirely independent of agricultural pursuits."[25] Salmon, who had the power to shape the BAI at its inception, won the initial rounds of this debate before the turn of the century, but his agency would continue to struggle with its split vision of social responsibility.

Cities of Animals

Everyone could agree on one thing: Americans' interdependence with burgeoning populations of domestic animals potentially ensured a high demand for veterinarians' professional services by the turn of the century. In response, veterinary leaders developed their profession to advise Americans on the social, biological, and economic management of their animals: what was safe to eat and how to keep an expensive horse healthy, for example. Ambitious veterinarians had every reason to believe that their professional scope would only enlarge with the expanding domestic animal economy. Nationwide, populations of animals had grown exponentially, especially during the last quarter of the nineteenth century. By 1900, 25.6 people occupied every square mile, on average, and so did 72.8 large domestic animals (not including dogs, cats, and poultry). Although they scarcely appear in most historical accounts, domestic animals were visible, necessary, and integrated into ordinary life at the end of the nineteenth century. Most Americans, whether they lived in urban, suburban, or rural areas, had some daily contact with horses, cattle, poultry, dogs, cats, or other animals that provided food, raw materials, and power for manufacturing, companionship, and transportation.[26]

With nearly three animals sharing the same unit of space with every human citizen, Americans' relationship with domestic creatures included a level of intimacy unknown to their great-grandchildren more than a century later. For rural people, animals' needs determined the day's routine. Since animals provided the motive power for transportation and farm work, the working day ended when they tired. Isolated rural children made pets of their animals and farm wives depended on dairying and poultry raising for the "pin money" that purchased household goods and sent children to school. Just about everyone assisted with the slaughtering and processing of food-producing animals every autumn, providing subsistence for the family. Fattening range-fed animals for sale to slaughterhouses and toil in the cities also proved a profitable enterprise for midwestern farmers. Altogether, the number of people involved in raising and caring for animals even part-time probably increased the official census figures for livestock raisers manyfold.[27]

Although they were usually raised in rural areas, large domestic animals were never far from urban markets and industrial appetites for raw materials. Animals and animal products passed back and forth easily over urban-rural boundaries, and penetrated every corner of industrial production because animals converted commodities into cash. On the great plains of the West, sheep and cattle made use of otherwise "worthless" prairie grass. On the farms of the Midwest and South, cattle and especially hogs thrived on acorns in the woods, scraps from the barn and table, fermented cut grasses and corn husks, and excess grain as well as feed grown specifically for them. These animals' bodies provided the raw material for many industries. Leather currying (curing) and tanning, harness and saddle making, wool milling, sausage grinding and meat packing, and trades like butchering and blacksmithing, employed one in eleven U.S. factory workers and tradespeople in 1890.[28] This statistic underestimated the importance of employment associated with animal industries in certain areas. In Chicago, for example, meat-packing companies employed 30,000 of the city's 295,000 workers, with an invested capital of over $66 million in 1900. Most lived in "Packingtown," the area immediately surrounding the stockyard lots (which Chicagoans proudly called "The Great Bovine City of the World").[29]

Even outside of Packingtown and other industrial areas, the mostly densely populated spaces for people also contained the most dense populations of animals in the United States. City dwellers saw, smelled, and benefited from animals every day. Dairy cattle, poultry, and pigs lived in backyards and barns; the stables and carriage houses that sheltered horses and the vehicles they drew lined many alleys. The concentrated numbers of these animals meant that they were a highly visible part of the urban landscape, perhaps even more so than rural landscapes. Because we associate animals with pastoralism, we tend to underestimate their urban presence a century ago. But as historian Philip M. Teigen has pointed out, a citizen was likely to encounter more horses in New Jersey than in Wyoming at the end of the nineteenth century.[30] This demographic observation challenges the representation of horses as primarily "home on the range," as immortalized in popular culture, and it prompts us to consider carefully the roles of animals in American cities.[31]

Despite their restricted physical spaces, cities contained large populations of animals because urban consumers were an end point of the market. Most cattle, swine, and sheep ended up in "cities of animals"—holding pens and stockyards—located at the edges of major urban areas. While many of these animals spent only a brief period in an urban stockyard on the way to slaughter, others, particularly milk producers, actually lived within the city for some time. In 1890, cattle, like horses and mules, were distributed throughout cities and suburbs of all sizes. Goats (kept for their milk) and sheep and swine (kept in stockyards or piggeries) could mostly be found in the larger cities.[32] Each of these types of animals served different purposes, exhibited unique distribution patterns, and presented different opportunities for the application of veterinarians' services.

As described by Charles Dickens, American pigs were the free-ranging "city scavengers" of the nineteenth century, but in the early twentieth century, the "republican pig," leading "a roving, gentlemanly, vagabond kind of life," had for the most part been corralled and relegated to city stockyards. Swine had been the first important raw material for meat packers in the mid-nineteenth-century development of the industry; decades later, swine populations were concentrated in the cities containing meat packers' facilities. In 1890, at any given time, one-third of Illinois' 150,000 swine could be found in Cook County, home of the stockyards on Chicago's South Side; 55,000 pigs were reported in Chicago in 1920. During these years, the percentage of all pigs in the United States living in nonrural areas doubled, reflecting the increasingly urban orientation of their consumers.[33] Pigs also found a ready home in and around cities as the recipients of edible municipal garbage because they transformed otherwise worthless refuse into meat, creating capital gains for their owners and assisting cities with the major problem of garbage disposal. Until piggeries' locations were restricted by many states after World War I, they were common within even large cities around the nation.[34]

Sheep, valued for their wool and meat, arrived in midwestern and western cities from the ranges where most of them had been raised. Most of Cook County's 24,000 sheep in 1900 were adult ewes, sent to the stockyards when they were no longer viable as breeding animals. By 1920, Chicago housed the single largest population of nonrural sheep in its stockyards, with Seattle and Denver not far behind. The distribution of the sheep population depended on the proximity of consumers and the workers needed to kill and process animals.[35] Beef cattle, especially adult steers, would also likely be found in city stockyards when they were not on farms or ranges. The numbers of nonrural beef cattle rose every decade, peaking with a 26 percent increase between 1910 and 1920. Most beef cattle were raised on the ranges of states such as Texas and shipped to feedlots for fattening or stockyards for slaughtering in and around Chicago, Kansas City, Omaha,

CITY OF ANIMALS: CATTLE AND SHEEP PENS AT THE DENVER UNION STOCKYARDS, 1920. *Courtesy of Western History Collection, "Stockyards, Union Stockyards, Pens," photo 18, Western History Department, Denver Public Library, Denver, Colorado.*

and other market cities. Chicago, the center of combined slaughtering and meat packing, ran 5.5 million beef cattle through its yards in 1890.[36]

People and animals in such high concentrations did not always live amicably together, and animals were to disappear from cities as part of the early twentieth-century rubric of urban improvement. Urban and periurban stockyards and packinghouses were located conveniently near manufacturing facilities that turned every part of the animal into some salable commodity.[37] Generally restricted to the poorest neighborhoods of cities, such as Packingtown in Chicago, noisome slaughterhouses and animal pens drew criticism from urban reformers for spreading waste throughout the city, creating "menaces to health." The Chicago River, winding through the city, transmitted the "filth from the stockyard sewer" discharged into its tributary, the Bubbly Branch. Caroline Hedger, a physician and settlement house worker, warned Chicagoans in 1906 that "if you are detained on the bridge that spans it, your clothing will smell for hours." The foul effects of animal nuisances also proliferated in human bodies. The "unhealthy" climate of Packingtown contributed to the poor health of workers, whose diseases could then propagate in the meat they packed and in the households of consumers. Hedger warned that even citizens not living in stockyard districts should be concerned "for their own sakes" about the health of the Packingtown residents.[38] Nor did the cattle, sheep, and swine living in stockyards create the only animal-related urban nuisances.

Dairy cows, too, were eventually slaughtered, but not until they had spent 2 to 10 years living in backyards, providing milk for individual families, or occupying large stables connected to dairies within town or city limits. Between 1900 and 1903, San Francisco had a dairy cow population of almost 6,000; one cow lived within the city limits of New Orleans for every 43 people at this time.[39] But cities were beginning campaigns to regulate dairy cattle and to remove them from within urban boundaries. The "backyard cow," an unwelcome occupant of increasingly valuable space and an unregulated source of milk, represented an urban nuisance. Beginning in 1902, Baltimore instituted laws requiring a certain amount of land for each cow; within four years, one-third of the city's cow stables (public and private) were gone. Buffalo, New York, adopted a similar strategy. Other cities, including Memphis, Columbus, Providence, and St. Paul, simply prohibited the keeping of cows within city limits. Dairy cow populations in large cities that had peaked in the first years of the twentieth century declined greatly between 1910 and 1920 with increasing urban regulation. San Francisco had outlawed most of its cows by 1920, and in New York and Washington, D.C., numbers plummeted 60 to 90 percent between 1910 and 1920.[40] The locus of dairy production shifted increasingly to suburban and rural districts farther away from the city centers; dairy cows' zones of habitation consistently increased in diameter between 1910 and 1930. The biological needs of large animals for food, space, and waste disposal threatened the order and efficiency that reformers sought to impose on the modern city.[41] The city was no longer a suitable place for these animals to live; it had become rationalized and sanitized.

Smaller animals more easily maintained their places in cities, towns, and suburbs in large numbers from the 1890s onward. Referring to the million-plus cat population of New York City, Edwin Tennery Brewster wrote in 1912 that the number of wild and tame cats together approached the number of voters. While no one knew exactly how many cats lived in urban areas, city dwellers agreed that the feline species was a highly successful urban commensal of humans. Both pet cats and strays ate table scraps, butcher's leavings, or other human food, and their small size and climbing ability made it easy for them to move around the congested city. Because cats were prolific breeders, their population in the springtime grew up to three times its yearly average level with the births of kittens. Most of the year, Brewster estimated, there were three cats in the city to each dog. Urban cats knew no class boundaries and were a ubiquitous presence in the lives of all; the legions of stray cats, often the unintended offspring of pets, became what Brewster called the "outlaws in the modern city." They proved themselves adaptable to the vagaries of city life over time and were therefore much harder to regulate than larger animals.[42]

Bigger and less resourceful than cats, dogs depended more upon humans to survive in the urban areas. People kept dogs for a variety of reasons: as pets, for security or sport, and as children's companions. Although a burglar alarm could replace a city watchdog, the urban citizen might find it more difficult to replace the intangible rewards of dog ownership. The family dog taught children important lessons about stewardship of dependent creatures, kindness, and fidelity. Their parents wanted dog company for the recreational and emotional advantages it imparted. Dog breeding and fancying, encouraged by the yearly round of dog shows, occupied the well-to-do. Sporting dogs made up perhaps the most unlikely canine group in urban areas, but "sporting" certainly could be defined broadly. Suburbanites and middle-class weekend hunters flushed rural fields, while their poorer urban counterparts promenaded with their dogs and pitted them against each other in races and fights. Police forces employed dogs around the city as "auxiliary policemen."[43] Canines remained prominent in cities despite attempts to regulate and tax them.

Finally, the most practically important and visible animals that towns and cities housed were large numbers of horses and mules who supplied the motive power for short-haul transportation. By 1910, 14 percent of the total horse population, and 6 percent of all mules, lived in nonrural areas. Two cities dominated the nation with their large equine populations: New York City, which alone contained 42 percent of New York State's urban horses; and New Orleans, in which 50 percent of Louisiana's nonrural mules lived. Their tremendous equine concentrations indicated that these cities, important as ports, also occupied regional positions of power as commercial hubs at the turn of the century.[44] Horses and mules delivered goods from the docks and warehouses to retail outlets in cities; they brought in crops and other raw materials to be processed; and they transported people on all but the longest journeys. Thus integrated into the urban economy, horses and their biological realities structured urban life in industrialized America.

Horses were essential to maintaining urban functions, but they (like other large animals) caused problems for the rationalizing, sanitizing city at the turn of the century. Residents of even fashionable districts in New York and New Orleans, pressing handkerchiefs to their noses, would hardly have been surprised to hear that their cities contained such dense populations of horses and mules. Well-to-do ladies crossing streets commonly paid boys stationed with brooms at the corners to precede them and sweep manure out of their way; manure piles at the sides of streets sometimes reached proportions so monumental that sidewalks were impassable. The bodies of horses that fell and had to be killed, or actually died while in harness, often lay unmoved for days, bloating and stinking by the sides of the streets. Urban residents worried that manure and rotting horse carcasses, long believed to be a source of disease-causing fumes or miasmas, also harbored the germs of tuberculosis and other diseases, which were thought to be spread by dust and flies.[45]

Indeed, in the late nineteenth century, the health problems common to domestic creatures and their human neighbors prompted city officials to consult with veterinarians (among other professionals) for expert advice on regulating urban animals. Outbreaks of rabies, also known as hydrophobia, had prompted periodic dog licensing campaigns because canines were blamed for transmitting the disease. In 1900, veterinary educator James Law espoused inspections, quarantines, and muzzling laws for all dogs, measures that he acknowledged would be strenuously opposed by dog owners.[46] Veterinarians, physicians, and health officers worried that urban cows carried diseases such as bovine tuberculosis, which could be transmitted through milk. High urban concentrations of pigs and garbage-feeding practices created an environment congenial to the spread of swine diseases such as trichinosis and hog cholera.[47] Such mounting problems concerned urban health officials, many of whom were physicians or political appointees with little animal experience.

Animals could also fail to fulfill their essential functions in cities, thus causing business and transport around the city to grind to a halt. City and town officials across the eastern United States had discovered this in late 1872, when the Great Epizootic, an epidemic of "equine catarrh," sickened large percentages of city horses. Officials blamed the illnesses of fire engine horses for the poor response of fire crews that allowed a conflagration to destroy Boston's downtown that year. Mass transit systems in cities farther west also broke down during the horse shortage; within 5 days of the disease reaching Cleveland, all streetcar service had halted. Glanders, a disease endemic in the crowded stables of city horses, was transmissible to humans and often fatal; if a public health official discovered the disease, the affected animals had to be killed and often the whole stable was quarantined. These diseases and others took their toll not only on bodies but also on profits. With every day that transportation was unavailable because of sick horses, business owners lost money and perhaps even their livelihood.[48]

Ambitious veterinary leaders, most of whom lived and worked in cities, saw all of these problems with urban domestic animals as an opportunity to develop a veterinary profession that could provide expert advice on how to manage animal health and illness. Their aspirations fit into a constellation of efforts aimed at regulating and improving social conditions in cities at the turn of the century. Veterinarians' goals paralleled those of engineers, planners, and others in maintaining what historians have called the "City Practical," the professionally managed city. Veterinarians perceived a need for their professional advice when progressive activists, interested in the conditions under which animal industry employees worked and lived, agitated for reform of stockyards and slaughterhouses. They could also help manage public health regulations concerned with dogs and dog waste in the interests of municipal hygiene and preventing rabies. Veterinary leaders believed that municipal regulation of the milk and meat supplies, combating flies, and other antidisease measures belonged within their province. Finally, veterinarians could keep urban transport functioning by treating horses' lamenesses and diseases. Thus, veterinarians interested in answering the call to social responsibility had plenty to concern them in turn-of-the-century cities in terms of both animal health and economics.[49]

Veterinarians were also concerned with regulation of urban animals because organized veterinary medicine was largely an urban profession at the turn of the century. In 1901, American Veterinary Medical Association president Tait Butler asserted that most veterinary school graduates earned their living as general practitioners, and 80 to 90 percent of them lived in urban areas containing more than twenty-five hundred people, not in the rural areas of animal production. In 1900 almost one-third of graduate practitioners lived in the nation's four largest cities—New York, Boston, Chicago, and Philadelphia. As late as 1915, "city-bred boys" dominated the population of graduate veterinarians, and most of those stayed in cities to practice.[50] This situation undoubtedly reflected many factors, including the national population distribution and the predominantly urban location of veterinary schools. These social structural variables, along with individual circumstances, certainly contributed to graduate veterinarians' preference for urban life. However, the most important underlying variable lay with veterinary practitioners' professional economic concerns. In short, they had to pay for their education and make a living, and large populations of valuable animals living in close proximity provided the best means of doing so.

Economic Necessity and Proprietary Knowledge

Cities harbored the highest density of America's most valuable animals. The economics associated with their function—especially in gratification of urban needs—determined the worth of many animals (and thus the extent to which their owners would spend money for their veterinary care). Cities excluded juvenile animals (which were useless as sources of food and power), who brought on average only one-fourth the price of mature animals and were unlikely to be the subjects of professional health care. Function and economics, however, often related to each other in a somewhat more complicated manner. Animals with desirable pedigrees, even though supplying the same utilitarian functions, brought more money than their less exalted conspecifics. The well-off individuals and businesses located in cities provided the largest market for pedigreed horses and their half-bred offspring. Geography also dictated value. Working animals living in the New England and Middle Atlantic regions brought the highest average prices in the country, helping to explain the concentrations of veterinarians and veterinary schools in Boston, New York, and Philadelphia.[51]

Veterinary schools reflected the importance of urban animals to veterinary practitioners by focusing their curricula on horses, the most valuable individual creatures. Private schools, located in cities and dependent upon student tuition, trained students largely as equine physicians, as veterinarian A. F. Schalk remembered. So did many of the university-associated schools: the University of Pennsylvania, located in Philadelphia, devoted a course to horseshoeing in 1906; Iowa State University, a rural school, included both horseshoeing and "hippology," or the study of equines, in its veterinary program. Anatomy and physiology, which traced their lineage directly to the eighteenth- and nineteenth-century French and British veterinary schools, had been largely constructed around the horse in American veterinary schools at the end of the nineteenth century. Every first-year veterinary student's course of study concentrated on the intricacies of an equine cadaver as the prototypical animal body. Although they sought to escape the old image of the uneducated blacksmith and horse doctor, early twentieth-century veterinary educators (among the profession's leaders) continued to focus on the equine species.[52]

Given these unpleasant cultural associations and the abundance of farm animal patients, the schools' equine focus and most graduate veterinarians' urban location seem paradoxical. Yet organized veterinary medicine's location as an urban profession depended most heavily on one feature: the location of dense populations of working horses in large cities. As historians Philip M. Teigen and Sheryl Blair have determined, numbers of veterinarians per square mile correlated most closely to those of horses in Massachusetts, for example; both were concentrated in Boston and its metropolitan area at the end of the nineteenth century.[53] This pattern reflected veterinarians' sage judgment of the animal marketplace on two counts. First, the densities of urban animals provided a critical mass of patient population base that "meant [economic] security for a veterinarian," as practitioner Arthur Goldhaft later explained.[54] Rural animals, spaced farther apart, could stretch a practitioner's time and profit margin to the breaking point.

Second, an intellectual and sociological focus on urban horses as patients represented a rational choice for veterinary practitioners because these were the most economically valuable individual animals in the United States. Adult mules and horses not only brought in more money for their owners than any other adult animal, on average, they also commanded value for their work. Businessmen could not afford to have their delivery horses unable to work because of illness or injury. As veterinarian A. H. Streeter had put it in 1897, "a sick animal means loss," and thus veterinarians could count on being paid to attend the horses and mules essential to urban commerce. Much of urban veterinary practice revolved around the treatment of lamenesses, common problems in hard-working urban horses. Many of the treatments characteristic of nineteenth-century veterinary practice—including corrective horseshoeing, external firing or blistering, and neurotomies—remained popular not only because of tradition but also because they were designed to preserve the animal's useful function (and thus its value).[55]

Thus, the developing veterinary profession in the late nineteenth century had decided to devote itself to caring for the commercial interests of animal owners; this was the most universally acknowledged articulation of veterinarians' social mission. Of course, different groups of veterinarians interpreted this mission in different ways and concentrated on different patient populations. Most veterinarians, who worked as general practitioners and concentrated on horses, saw their own prosperity as part of a larger common good. The economic panics of the 1890s, in which horse values plummeted, drove this lesson home to struggling veterinary practitioners. By 1900, however, horse prices and urban demand for horse power had increased again. When practitioner Arthur Goldhaft got his veterinary degree from the University of Pennsylvania in 1910, he immediately opened an office in Philadelphia on Pine Street's "Doctor's Row." His practice grew slowly at first, although hundreds of potential equine patients trotted past his office daily. His big break came when, through old neighborhood connections, Goldhaft landed a lucrative contract caring for the large numbers of horses housed in a nearby livery stable. He spent his days combating colic, infectious diseases, and lameness problems; his steady job made him feel as though "the future looked great." Like most urban veterinary practitioners, Goldhaft saw himself primarily as an equine physician, and he and his urban colleagues represented the largest single group in the young profession.[56]

While urban practitioners concentrated on horses, their less numerous colleagues in rural areas cared for cattle and other animals as well, and a small group of veterinary leaders used the umbrella of commercial interest to cover work in public health, scientific research, education, or administration. Veterinary researchers at the BAI focused most of their experimental investigations on economically valuable groups of food-producing animals. Indeed, epizootics of hog cholera, Texas cattle fever, and bovine pleuropneumonia were paramount among the BAI's concerns before 1900. Epizootics encouraged livestock producers and the U.S. government to employ BAI veterinarians as researchers and eradication officers because these diseases struck what veterinarian William H. Lowe called "the most sensitive thing in all creation—the bank account."[57] Of course, employing veterinarians in response to any animal health problem implied a certain faith in their ability to control disease and restore health. Before the 1880s, veterinary consultants' claims to expertise and efficacy rested on their European educations in the established veterinary schools of Edinburgh; London; Alfort, France; and Berlin, which instilled in them an interest in scientific investigation as well as eradication of disease.

Indeed, the migration of European veterinarians to the United States, livestock disease problems, and the development of germ ideas all converged in the 1870s and early 1880s to create the prototypical modern American animal expert. The early leaders of the American profession had been trained in various traditions, but they were all self-defined by their interest in not only eradicating diseases but also understanding pathological mechanisms.[58] These veterinarians' credentials (many had trained in the laboratories of Louis Pasteur, Joseph Lister, and Robert Koch) dictated their beliefs about how disease arose and took hold in an animal's body. Perhaps the most influential proponent of the developing "germs theory" in American veterinary medicine was French expatriate Alexandre Liautard, through his editorship of the *American Veterinary Review*. Beginning with the *Review*'s first volume in 1877, Liautard published the literature of his homeland and other European nations in translation, along with American articles and case reports. A firm proponent of Louis Pasteur's work, Liautard brought the particulars of the germ theory debates to U.S. veterinarians (the *Review* was apparently widely read—it was quite a successful business venture).[59] Although not without opposition, the firm standpoint of the *Review*'s editor and his like-minded colleagues steered veterinary education and the developing veterinary profession toward experimental medicine and the laboratory sciences in the 1870s and 1880s.

Simultaneously, employment of these "scientific" veterinarians by the U.S. Department of Agriculture also played a leading role in boosting the importance of the laboratory sciences in veterinary education and regulatory matters. Bacteriology, pathology, and physiology—all components of the new experimental medicine—had several advocates among graduate veterinarians. Scotsman James Law, educated at Dick's veterinary school in Edinburgh, brought his ideas on contagion with him when he arrived in the United States in 1868 to found the veterinary program at Cornell University.[60] Law had an immense influence on the development of veterinary education and federally sponsored veterinary research. Cornell's program was the most widely emulated model of twentieth-century veterinary education. Law provided expert advice for the USDA in the years prior to the establishment of the Bureau of Animal Industry, and his students controlled the BAI for the next 30 years.[61]

With the chartering of the BAI in 1884, these veterinary scientists attained the resources necessary for conducting research that they argued would protect the U.S. livestock economy. As a federally funded institution, the BAI owed its existence to a specific set of biological, economic, and political circumstances. In the 1870s and 1880s, food animals and meat products exported by the United States began to encounter hostile receptions in Europe. Germany and other nations with state-inspected meat-producing industries protested that uninspected American products did not meet the recipients' high regulatory standards, and U.S. exports declined dramatically. Moreover, in the late 1870s, epizootics of bovine pleuropneumonia and other "animal plagues" led to great economic losses for American cattlemen at home.[62] Faced with livestock owners clamoring for a remedy to animal disease problems, Congress authorized the Treasury Department to create a cattle commission and build quarantine stations in 1881 and then, 3 years later, transferred this work to the USDA and created the BAI. For the next 70 years, BAI scientists investigated and isolated outbreaks of animal disease, conducted basic disease research, and secured inspection and quarantine stations for import and export of animals and meat products.[63]

James Law's former student, Daniel E. Salmon, as chief of the BAI began a program of research that depended on annual renewal of federal appropriations. Salmon set up the bureau and chose particular animal disease problems in order to accomplish several concurrent goals. Foremost, bureau veterinarians and scientists needed to assert that their work helped to protect the economic value of stock owners' living property. From the BAI's inception, Salmon (and others) put the value of research in terms of the millions of dollars that could be saved if certain animal diseases could be eradicated.[64] As a research and regulatory organization, the BAI's official (as well as historical) raison d'être and its claims to federal support depended upon the economic value of cattle, sheep, swine, and other livestock. This economic motivation combined rather well with another goal, the extension of laboratory methods into practical applications. In accordance with his own education and methods, Salmon arranged research groups at the BAI around specific laboratory and field experiments designed to reinforce and extend the intellectual development of germ theories and practices. Along with leasing land for field work, Salmon scrounged equipment for pathology and bacteriology laboratories. He hired scientifically trained staff, many of them Law's former Cornell students, including veterinarian Fred L. Kilborne as the director of the field experiment station. Another Cornell-trained scientist and physician, Theobald Smith, began his career in 1885 as the director of the pathology laboratory.

Kilborne, Smith, and another BAI veterinarian, Cooper Curtice, delivered the BAI's major scientific achievement (which also could be touted as an economic success) while working on Texas cattle fever in the 1880s and 1890s. This disease prevented the transit of cattle from the southern to the northern United States, a tremendous problem given the concentration of markets, meat packing, and meat consumption in northern cities. Endemic but usually not fatal to southern cattle, it killed northern cattle exposed to their southern counterparts with such frequency that vigilante border patrols had sprung up in Kansas, Missouri, Illinois, and other states to prevent bovine transit from the South. Daniel Salmon had mapped the peculiar geographic characteristics of the disease in the 1880s, then turned the problem over to Smith, Kilborne, and Curtice. The details of their intensive 5 years spent on the experiments make for interesting analysis on a number of levels, but most obviously because their conclusions literally added a new dimension to the international framework of germ ideas. They proved that a vector organism, in this case a species of tick, could be necessary to pass a germ disease from one animal to another. Their results influenced research being done on a number of previously mysterious diseases, including malaria, that required vectors (often insects) for the transmission of disease in animals and humans.[65]

Smith and Kilborne's 1893 report, published as a special *Bulletin* of the BAI, simultaneously made a revolutionary contribution to the experimental medical sciences, provided a mechanism for controlling the disease, and promoted BAI research as useful to the health of the livestock economy.[66] Salmon directed his scientists to continue research on a vaccine against Texas cattle fever and eventually supported efforts to dip cattle in a tick-killing solution so that northern livestock markets could be opened to southern animals.[67] Salmon's strategy of dependence on basic research, in retrospect so successful, represented a gamble at the time. Only a few years after Robert Koch's famous demonstration of the bacillus that caused tuberculosis, the scientific community had still not come to consensus on the particulars of the germ theory. More important, another decade would pass before experimental work produced even a limited number of therapeutic or preventive applications of any importance to either animal or human health.[68]

Not all veterinarians supported Salmon's intellectual mission at the BAI in the 1890s. Many veterinarians, like their counterparts in human medicine, viewed the idea that states of health and disease could be controlled by microscopic germs with suspicion. Such a reductionist model threatened traditional treatments and patterns of veterinary authority, which were based on experience rather than experiment. Some practitioners' reluctance to abandon prevailing procedures, coupled with few demonstrable applicable benefits arising from experimental medicine, helped to maintain a certain distance between the BAI and the average veterinary practitioner until the turn of the twentieth century.[69] Even veterinary scientists in sympathy with experimental methods caused problems for the fledgling BAI. The most ardent competitor was the German-trained F. S. Billings, based at the Nebraska Experiment Station. Billings' claims to have discovered the "true germs" causing diseases such as Texas cattle fever and swine plague directly challenged the findings of BAI scientists, who were perhaps forced to document and confirm their results more carefully in defense.[70]

These debates reflected the equally valid but conflicting positions of veterinary leaders and researchers on the larger social purpose of veterinary medicine. Could all of veterinary medicine's missions be included under the umbrella of economics? The BAI was an exemplar of the profession's identity issues, finding itself caught in the middle of its constituencies' conflicting requirements. Its regulatory functions (particularly meat inspection) occasionally sacrificed food producers' profits to the tenets of public health, and its research programs often failed to deliver on their promises of beneficial solutions to the problems of livestock disease. Nonetheless, the BAI continued to receive congressional appropriations based on BAI-employed veterinarians' reliance on the idea of "science" as a claim to competence. Since "science" was a plastic concept that meant different things to different constituents, Salmon and other veterinarians could invoke it as a methodology likely to satisfy those concerned with the livestock economy as well as public health.[71]

Although the percentage of scientifically trained men in the veterinary profession was increasing, the BAI's cadre of veterinarians did not represent the majority of the profession, whose livelihood still lay with horse doctoring. In 1900, veterinary leaders and researchers served the national interests of the livestock economy while market forces determined the work of local independent practitioners. The BAI concerned itself with large populations of animals, usually those raised for food; when they were unable to find curative therapeutics for animal diseases, BAI veterinarians concentrated on preventive medicine. Most veterinary practitioners around the nation, however, still treated the injuries and illnesses of individual animals (usually horses), with little attention to prevention or to large herds (unless they were faced with a local epidemic disease; or had a rare contract with a dairy, ranch, or other business). Their interest in forming a profession stemmed primarily from a monopolistic goal of shutting unlicensed competitors out of the market, whereas veterinary leaders and researchers hoped to use the laboratory sciences as a basis for obtaining proprietary knowledge and establishing veterinarians' social role as animal experts.

By 1900, the two groups did maintain tenuous connections through the programs of the BAI. Beginning in the 1890s, BAI scientists produced quantities of mallein (used to test for glanders in horses), tuberculin (for tuberculosis in cattle), and other diagnostic testing agents, and they distributed them to veterinarians around the nation free of charge. Outbreaks of highly infectious and contagious animal diseases also tended to unite bureau officials with local practitioners. In the 1920s, BAI scientist Ulysses G. Houck remembered the cooperation of privately practicing veterinarians around the nation who reported and helped to combat episodes of foot-and-mouth disease in 1902, 1908, and 1914. The 1914 outbreak, which started in the Midwest and spread through twenty states, subsided with the united efforts of about 450 BAI inspectors and nearly as many state employees and private practitioners.[72]

At that same time, the BAI began surveying and accrediting veterinary schools around the nation, which proponents saw as "elevating the standards of the veterinary profession in this country," but which critics viewed as unwelcome federal interference in the autonomy of private and university-affiliated schools. "The federal government furnishes not one penny for veterinary education," complained Leonard Pearson, dean at the University of Pennsylvania, "yet is every year taking, without return, the results of veterinary teaching."[73] Those "results," trained students, could only sit for the BAI civil service examination if they had attended an accredited school. The concerns of veterinary educators about state control did not dampen veterinary students' eagerness to be eligible for BAI jobs, since such employment was far less economically risky than establishing a private practice. Certain veterinary schools' ability to attract students thus depended on BAI accreditation and many chafed at the BAI's increasing authority over veterinary education.

Greater influences than the BAI, however, promised to restructure veterinary practice and the veterinary sciences. Ever-changing scientific ideas, the challenges of modern life, and (most crucially) changes in animals' roles in American society affected all veterinarians, whether research scientists or private practitioners. At the end of the nineteenth century, most veterinarians were passive provincialists, maintaining their individual practices. They felt little need to become actively engaged with the larger transformations in social conditions that surrounded them. The days of such complacency were numbered, however. Back in Philadelphia, the young and optimistic practitioner Arthur Goldhaft began to feel uneasy about his prospects within 5 years of establishing his practice. Both his own feelings and external factors prompted him to think about leaving his lucrative practice and livery stable job. Disgusted by the sordidness of Philadelphia's corrupt political system and the vices of the city, Goldhaft's weekends in the countryside with his family felt "like a bath." He had also noticed that in 1912, "the tin lizzies were showing up on every street." Soon after, the closure of a large Philadelphia livery stable shocked Goldhaft into reevaluating his professional situation. "What would a horse doctor be without horses?" he wondered. He and other urban veterinary practitioners saw "the handwriting on the wall" that their roles in the traditional horse-powered economy were threatened. As the value of horses fell, so would the need for veterinary services. What would happen then? How quickly would it happen? Taking no chances, Goldhaft moved with his family to the New Jersey countryside, starting a new life as a country veterinarian away from the horse-rich city he had known.[74]

CHAPTER TWO

Valuable Patients

Horses and the

Domestic Animal Economy

Arthur Goldhaft's pessimism about the importance of urban horses lay far in the future on the morning of May 30, 1903, when Bostonians gathered for their first annual workhorse parade. Hundreds of working horses, hitched to the wagons they hauled every day, paraded past thousands of people lining Commonwealth Avenue and Beacon Street. Coats shining, manes combed, harness cleaned, with entry numbers gently flapping on their collars, the horses stepped out lightly and smartly in front of excited drivers and empty wagons. Judges awarded ribbons and prize money "for good, hardworking condition, docile and gentle manners, and for comfortable harnessing" to the best horse and driver in each class. Grocers' teams, truckmen, milk haulers, and express deliverymen who passed unnoticed during the week stood out on this Memorial Day morning as proud representatives of civic virtue.[1]

The purpose of the parade, explained Henry C. Merwin, president of the Boston Work Horse Parade Association, was one of social reform: "to make the public generally interested in the horses which they see at work every day in the streets . . . to think more of the welfare of horses." He added that the parades sought to educate drivers on proper feelings toward and treatment of their horses, and to avoid vices such as alcohol consumption. Merwin stressed that the Boston parade, unlike its highbrow predecessors in England, aimed to give "poor horse owners every possible opportunity" to shine by charging no admission fee, giving prizes for the best old horses and driver-owned horses, and discouraging the use of new or fancy harnesses and wagons. At the same time, the parades encouraged all to "buy and use horses of a fine type." Raising the standard of the city's working horses, Merwin argued, would mean that the animals would "go through life with less suffering than the inferior, coarse-bred horse." By all accounts, the Boston parades were a tremendous success; by 1907, at least fifteen other cities in the country held parades on the Boston model. Philadelphia's parade in that year drew more than a thousand entries and throngs of spectators; New York's boasted 1,371 horses in the lineup.[2]

Workhorse parades, like all cultural spectacles, teach us much about the social and cultural issues surrounding their subjects—working horses and their working-class drivers—in the 1900s and 1910s. At once devices used by middle-class activists to impose humane education on workers, and emblems of working-class pride, these parades personified the contradictions implicit in social reform efforts at this time.[3] They also demonstrated the ubiquity of dependence on horses among all classes and races of people, while allowing organizers to conceal class-based reform ideas within their stated purpose of humanitarian education. Even though the drivers, horses, and wagons featured in the parades were distinctly modest in means and appearance, everyone knew that those whom judges dismissed as "coarse-bred" horses and owners would not win any prizes. Yet workhorse parades were also different from most social reform efforts in an important way: humane reformers' stated primary concern, while shaped by classist and racist assumptions, lay with the welfare of the animals. To them, horses were fit subjects of humane attention within the framework of the animals' accustomed social role. In no way did humane reformers champion a prohibition on the use of horses for transportation and hauling goods; rather, they sought to create a public image of the proper type of use—that which supplied human necessities while elevating both human sensibilities and the animals' quality of life.[4]

Veterinarians' professional culture had traditionally stressed an unsentimental attitude toward their patients, but that did not prevent them from cooperating with events such as workhorse parades that promised public exposure and approbation for the veterinary profession. Leonard Pearson, dean of the University of Pennsylvania Veterinary School, chaired the Pennsylvania Work Horse Parade Association and filled the organizing committee with other veterinarians in 1907. By this time, veterinarians had built casual alliances with humane organizations. State laws often required the oversight of animal cruelty cases by both animal welfare officers and veterinarians, and so the two groups knew each other personally.[5] Thus Pearson's cooperation with humane reformers on the workhorse parade in his state did not surprise the editors of the *American Veterinary Review,* who reported on the event and praised the Pennsylvania veterinarians' participation. Most veterinarians in the early years of the century maintained private practices that focused on horses as patients, and public roles such as Pearson's encouraged Americans to identify the veterinary profession closely with horses as well. Workhorse parades continued into the 1920s, but by that time they had diminished significantly in size and importance; there weren't enough horses left in cities to put on a good parade, and few spectators were interested in watching them. Working horses were becoming an anachronism, both practically and culturally, as trucks, automobiles, and electric streetcars multiplied on city streets.[6]

The replacement of horses with motorized vehicles began with the wide use of mechanized urban streetcar systems in the 1890s and continued through the 1940s as farmers purchased tractors and trucks. It is difficult to overestimate the impact of this transition away from horse-powered society. Life in America had been structured around the pace and needs of horses, the "living engines" of commerce, agriculture, and recreation. Horses' biological abilities determined how far and how fast one could travel, the circumference of cities and their suburbs, and the rhythms of farm life. Horses had long been cultural and social icons as well as crucial economic resources; this meant that Americans' attitudes toward horses both as living beings and as technologies influenced the manner in which horses maintained or lost economic value and social relevance. Thus, the transition from horse to motorized power proceeded unevenly and slowly, but it definitely did include tremendous decreases in horse populations and economic value nationwide, especially between 1910 and 1920. This basic fact had far-reaching ramifications, particularly in urban areas. Employment patterns changed, for example; in 1910 about 3 percent of the total workforce in the United States owed their jobs to the equine economy, and they presumably lost work or income with its decline over the next four decades.[7]

Certainly veterinarians viewed the eclipsing of their "most profitable patient" with alarm, and the declining use of horses proved to be a crucible in which the young profession would have to reform itself. Veterinarians' reliance on the horse economy put them in danger of losing an important professional identity, mission, and economic niche between 1900 and the 1920s as horse populations and values declined. Veterinary leaders and practitioners responded to this challenge by adopting a variety of strategies that eventually unified the profession's concentration on the needs of food-producing animals and public health. In doing so, veterinarians acted on a sage assessment of which animals Americans valued most and for which they were most likely to demand professional expertise. The transition away from horse power taught veterinarians the importance of animal value in determining their strategies for future professional success. At the same time, their leaders rearticulated the profession's social and moral responsibilities to reflect the new realities of the domestic animal economy.[8]

Classifying Horses and Their Owners

Back in 1900, veterinary concentration on horses was justified, since these animals figured prominently in American life as both practical necessities and social badges. Their value arose in part from the fact that class distinctions applied not only to people but also to animals. Like automobiles a century later, horses came in different sizes, were developed for particular purposes, and appealed to different classes of buyers. Working people's horses, stolidly drawing wagons and carts, lacked the style of an upper-class lady's carriage horse. Nor could the fine-boned legs of carriage horses bear for long the heavy work of pulling streetcars or agricultural implements. Individual horses occupied positions in a hierarchy based on their characteristics, their occupation, and the social status of their owners; all of these attributes contributed to a horse's value.[9]

Horses also advertised their owners' social status. Citizens used their horses and conveyances to reinforce their social distinctions publicly. The attributes of horse-drawn vehicles were remarkably complex indicators of social place. Undue ostentation or improper outfitting of equipages marked the owner as a vulgar parvenu and such public display was a language understood by everyone from socialites to servants.[10] This visual language of classification applied not only to humans in horse culture but also to the horses themselves. Owners prized horses whose action, color, and especially speed could set them apart: as the *New York Herald* explained in 1893, "to have money and not own fast horses is to be nobody." In the last three decades of the nineteenth century, the "fast set" was just that, dashing about in their gigs. Although anyone with money, no matter how ill obtained, could buy the necessary carriages and horses, some differentiation could be maintained between the nouveau riche and their social (if not economic) betters through the elaborate protocols and specifications for equipages. The amount of brass on the harness, the horse's color and markings, and the heights of footmen were among the symbols that conveyed the tastes and social position of the patron. Such protocols, along with the expense, also served to keep professionals, merchants, and other inferiors out of the carriage parade.[11]

The middling sorts had their horses, too, however: the better cart and delivery horses, hacks, farm horses, and other moderately respectable driving horses maintained their place in the horse hierarchy by reflecting the ideals and wholesome characteristics of their owners' social position. American notions of thrift, hard work, and (limited) social mobility found their expression in plain, "honest" horses. Breeds that had originated in the United States, such as the Morgan, drew special praise as national symbols whose virtues included "great courage, loyalty, and a perfectly temperate disposition." Morgans were the preferred driving horses of country physicians and veterinarians; although they were not the fastest or flashiest animals, they trotted reliably all day and their drivers could depend upon their "smart and sensible" natures. Such attributes were carefully considered by potential owners, because a horse and buggy represented a serious investment for the small business owner at the turn of the century. The total initial cost was around $490, and a horse and buggy required $350 per year in upkeep against an average professional salary of about $600 per annum. Nonetheless, possession of a horse was essential to a geographically widespread practice (and often to a socially successful one).[12]

Equally important to their owners were farm horses and their urban counterparts, the better class of cart and delivery horses. At least symbolically descended from the European warhorse, the light-draft animal drew the plow that broke the plains and was also a "fit servant of commerce" in the city. "It is in the city that we find the cart-horse in his noblest form," wrote H. C. Merwin, the Boston workhorse parade organizer. A "true son of the soil," the "picturesque" cart horse brought a little of the "glory of nature" into the city. The strength and natural splendor of such a horse "must interest everybody who cares for the beautiful or picturesque." Likewise, the horses that drew fire engines at top speed through city streets commanded admiration. In accordance with his social position, the fire horse's life featured "good oats and hay, a clean warm stable and comfortable bed, elaborate grooming and gentle treatment." With his "companions, brute and human," explained Merwin, "his social affections and faculties are highly cultivated." Although not necessarily of special breeding, a horse working out of a fire house received the good treatment to which his occupational status entitled him.[13]

Worn-out cart horses, fire horses, and delivery animals often ended up at the bottom of the horse hierarchy, drawing streetcars or carts owned by impoverished people.[14] Certainly, many of these horse owners did their best to care for their animals (to protect their investment, if for no other reason), but their resources were too limited to afford the better, $30-a-month livery stables. The misery of the horse's life often reflected that of his owner: "[the horses] have scanty fare, little or no clothing, hard boards to lie on, and, commonly, severe toil to endure."[15] Humane associations improved the lot of a few horses, donating blankets, bedding, feed, and veterinary care to "deserving cases" and running farms for equine pensioners. One driver received a new horse blanket when he admitted that he only had a ragged old quilt to cover his horse, and he thought that naughty boys pelted him and his horse with snowballs in the winter because of his horse's shabby clothing.[16] The boys recognized the man's social vulnerability—he collected and resold trash for a living and spoke broken English—by the appearance of his horse, underscoring the importance of horses as visual symbols of their owners' class and cultural characteristics in the early 1900s.

The Transition Away from Horse Power

Horses' complex symbolic value meant that when alternatives to horse power became available, animal owners did not inevitably or immediately adopt the alternatives because of their perceived technological superiority. Bicycles for individual transportation and electric streetcars for mass transit appeared in the 1890s; trucks and automobiles first came into use in the 1900s. Automobiles became widely available with the introduction of the Model T Ford in 1908, and tractors likewise in the 1920s. Faced with new choices of steam, electric, or hydrocarbon-based power, millions of Americans weighed the advantages and disadvantages of horse power and motor power. Their calculations illustrated not only practical considerations but also complex social, emotional, and cultural ties between horses and humans. Those Americans who preferred to keep their horses often cited the difficulties of restructuring work or travel practices, or the affection they felt for their animals. Turn-of-the-century cultural currents also greatly influenced these decisions, as horse supporters and advocates of motor power challenged each other in debates over the meaning of "modern" and "progress."[17] The transition in motive power proceeded, then, incrementally through the millions of decisions Americans in urban, suburban, and rural areas made about how they valued their horses.

Americans' careful evaluation of working horses and motorized alternatives was in itself what historian Ronald R. Kline has called "a means of social transformation"; those needing power decided how horses or motorized vehicles would fit into or alter their lives and daily tasks.[18] Many Americans associated horses with being old-fashioned and antimodern and devalued them accordingly. Even before the turn of the century, the horse had become a symbol of antimodernism, in the technological sense. The editors of *Scientific American*, unabashed technological boosters, had predicted in 1894 that "the tendency of the present day is that the horse must go, must go metaphorically, for his days of labor seem nearly passed. Verily, the field of usefulness formerly held by the horse is narrowing daily."[19] Over the next 20 years, horses' cultural as well as technological value suffered in comparison with that of motorized vehicles. In 1914, veterinary educator Carl Gay complained that the technological promise of motorized vehicles was exaggerated because they were "in vogue." "The motor idea is conspicuous in our mode of dress, our conversation, [and the dreams of] small boys," he wrote; the cultural interest of Americans defined the 1910s as the "motor age."[20] As the harbinger of modern life, the automobile not only carried Woodrow Wilson to his 1913 inauguration, it also reconfigured cities, suburbs, and life in rural America. By World War I, Americans valued the automobile as the enlarger of lives, expander of horizons, and means to understanding and living in the modern world. "Think of what it means," insisted writer Henry Norman. "It is a revolution in daily life. With an automobile, one lives three times as much in the same span of years, and one's life, therefore, becomes to that extent wider and more interesting." By annihilating time and space, the automobile helped to redefine the scope of living in the modern world; it symbolized progress, leaving horse power as an anachronism.[21]

These were not the only definitions of progress, however. Other Americans viewed the question of modernity in the new century as a moral problem, and their arguments often (but not always) championed the continued use of horses. Horse supporters cited the importance of "horse sense," the "noble horse," society's duty to "man's best friend," and the importance of a human-animal bond in an increasingly mechanized world. Horses could be valued as "friends" of their owners; "I like my horse in a way that I could never like an inanimate object like an automobile," wrote F. H. Moore in 1910.[22] To horses were attributed the "sense" that ensured the safety of their drivers. "The horse used to bring the worn-out reveler and the tired doctor home," explained Joseph K. Hart in 1925. "Now the auto, its steering wheel slipping from exhausted hands, flings them both into the telegraph pole and leaves their broken bodies as warnings that a mechanism does not supply its own intelligence."[23] Some commentators encouraged the preservation of horses in daily urban life because horses represented nature, a necessary remedy for the stresses of modern life. "The neigh of a horse will always appeal more strongly to humanity than the honk of an automobile horn," an Iowa newspaper editorialized.[24] Especially in cities, the loss of animals, an important connection to the morality of nature and the agrarian past, alarmed citizens. The modern urbanite would suffer a loss of "finer feelings"; as writer Rene Laidlaw asserted, "we can never grow to *love* the man-made inventions."[25] According to Laidlaw, these considerations affected people's choice of horse or motor power, as well: "emotion . . . explains why some of us still cling to the horse, figuratively, and . . . literally too." Horse ownership helped to preserve human sensibilities in a world increasingly dominated by nonhuman machines.[26]

The associations of motorized vehicles with technological progress and horses as sentimental favorites occasionally got overturned, however. For example, horses remained the practical choice for many occupations well into the twentieth century. Fire horses endured largely because horse-drawn fire equipment was difficult to convert to motor power, and the new motor equipment cost too much. In 1912, more than thirty thousand horse-drawn fire wagons were still in use in the United States. Horses found other niches as street sweepers and short-distance haulers. In 1924, the city of Philadelphia spent $108,000 keeping its seventeen hundred street-sweeping horses in shoes. Retail grocers, who still picked up and delivered small amounts of goods over short distances, preferred the cheaper and more familiar horse and wagon. "Compared with motor cars," wrote one grocer in 1922, "our horse equipment is one-half cheaper to buy, lasts twice as long, and gives more dependable service under all conditions." Dependable service meant in part that drivers' familiar daily rounds, structured by the pace of the horse, would not have to be changed.[27]

On the farm, too, the structure of days, seasons, and years had been built around the patterns of horse power, and some farmers were reluctant to reconfigure their daily routine. Owing to cost and unreliability, many farmers with small to medium-sized holdings did not acquire a tractor until around World War II. Those who did usually farmed with both animal and mechanical power for many years. Most of their horses worked in fields, especially on small farms, and on any farm that grew corn. This crop benefited from careful cultivation and the avoidance of soil compaction, both attributes of horse work. Horses also excelled at short-haul work and were only slowly replaced by tractors and trucks. Fewer than one-third of all midwestern farmers owned a motor truck in 1930, regardless of whether they owned a tractor. A 1929 study of Minnesota farms revealed that two-thirds of even the most prosperous farmers did not own trucks (and 50 percent did not own a tractor).[28] Although horse numbers were certainly decreasing in rural areas, farmers still valued them for particular jobs. Farms and ranches were the last bastion of animal power, owing to economics but also in part because removing horses meant upsetting the entire structure of farm work.[29]

Farmers consulted more than practical issues when deciding whether to mechanize. Some disliked horses; others valued their relationship with their animals and acted accordingly. As one farmer put it, "I have a member of the Old Dobbin family that is of the third generation which, together with his two direct forbearers, have rendered constant and unrequited service in our family for practically 60 years—and the possibility of not continuing that co-partnership presents a problem I just don't want to contemplate." Of course, many farmers and farm tenants kept working with horses and mules because they could not afford mechanization. The fact that even some well-off farmers were reluctant to give up their draft horses led to tractor advertisements that asked them "Have you placed a sentimental value on your horses out of proportion to the work they are able to perform?" Some farmers continued to value horses, even idle ones, well into the 1930s. In 1933 census officials wrote that sentiment could not be overlooked in the study of the horse situation. Their analysis revealed a correlation between the age distribution of farmers and farm horses in New England, which officials attributed to elderly farmers retaining the horses until they died of old age. The same striking pattern was found in every state. "The very long retention of older horses" was characteristic of farmers at least 45 years of age, even after they had acquired tractors, trucks, or automobiles.[30] Although the census officials who wrote this report did not analyze these findings further, farmers' patterns of horse retention could be read several ways. Census officials implied that elderly farmers identified with the plight of elderly horses, perhaps finding expression for their own thoughts about the end of life. Moreover, these farmers' valuation of their animals may be interpreted as a response to the uncertainties of interwar transformations in American society (of which motorized technologies were an important part); keeping aged, useless horses was an economically unfounded but morally significant response to modernity.[31]

However, morality could also be used to make a case against the continuing use of horses as workers. As that argument went, horses and mules had offered "unswerving faithfulness, unquestioning devotion and unmatched achievements in blazing the way for the progress of the world," and humankind owed them consideration in return.[32] Many horse lovers urged the replacement of horses with motor power and the animals' release from the unrelenting labor they had performed in cities. The "toilers in the traces" suffered beatings, hard labor, and exposure to the elements. An *Independent* editorial described a typical winter day in New York this way: "The horses flounder, straining at the tugs, their mouths frothing at the irritating jerks of the bits, the skin on their flanks quivering under the repeated strokes of the whip-lash in the hands of the angry driver. . . . Here is one which has given up and laid down on its side, preferring to die in the soft snow bed rather than keep on." The coming of the motor vehicle, "which never gets discouraged and does not feel a blow," would free horses and mules "from servitude. . . . The burden of traffic will no longer be laid on flesh and blood."[33] Even horse-supporting editors of the *Rider and Driver* proclaimed themselves "glad to see [working] horses gradually being replaced . . . as that is relieving the poor beast of cruel drudgery." The *Rider and Driver* could happily turn its attention to horses of "sport, pleasure and fashion" and provide rhetorical support to humane societies and other organizations concerned with the quality of the remaining working horses' lives.[34]

Humane associations and societies for the protection of animals in the United States had been established in part to address the abuse of horses. Along with sponsoring workhorse parades and public watering troughs,

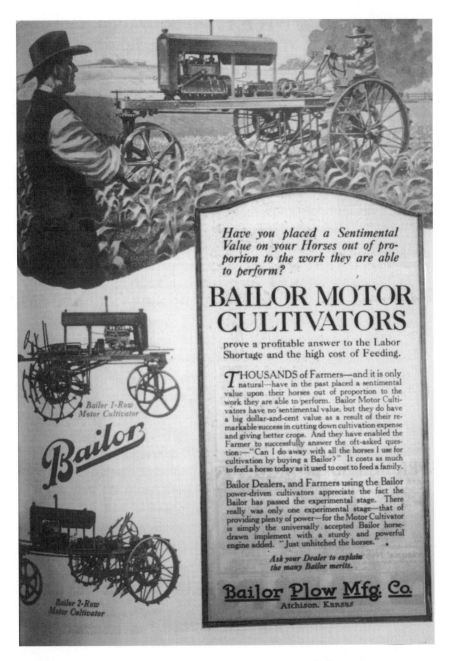

MACHINERY MANUFACTURERS RECOGNIZED HORSES' "SENTIMENTAL VALUE" AS A FACTOR IN FARMERS' PURCHASING DECISIONS. *Bailor, Inc., advertisement, from the* Country Gentleman *85 (March 27, 1920): 71.*

these organizations supported the development of technologies to replace horses in heavy work. The most miserable occupation for a horse was pulling streetcars: "the heaving flanks, the tortured mouth, the nervous eye, of the car horse,—the excruciating sound of his iron-shod hoofs slipping and clashing over the pavement in a vain attempt to start a heavy load." Beginning in the 1890s, animal protection activists had rejoiced that the car horse was fast disappearing, replaced by mechanically powered streetcars. "The electric current that invisibly takes his place has not capacity for suffering," wrote H. C. Merwin, "and [the car horse] will thus be released by Science." The release of the streetcar horses from such cruelty provided a key benefit to humankind, also. As the editors of the *Independent* wrote in 1914, "In freeing the animals from the evils of slavery we are freeing man from the evils of mastery." This evocative language could not have failed to recall arguments for the abolition of human slavery 50 years earlier. Efforts on behalf of working horses defined humanitarianism very broadly as a saving grace for all of civilized society, for humankind as well as horse-kind.[35]

These humanitarian arguments, in all of their iterations, provide a sense of how progress, technology, and modernity became entangled with humanity, cruelty, and sentimentality in the opening decades of the twentieth century. Farmers, tradesmen, city officials, deliverymen, and commuters factored many variables into the calculus of how they valued horses and machines at the turn of the century. Although they were more powerful and faster than horses, machines could be expensive and unreliable. Machines opened the door to a new way of life, but they lacked the emotional satisfaction of the human-animal bond. The traditional visual language of class encoded in a properly equipped horse and carriage could be transferred to automobiles, but it would take time for the new distinctions to become established and widely understood. These concerns meant that at least early in the century, the choice between horse power and machine power was hardly obvious or predetermined, nor was either irrevocably associated with liberation and progress. Nevertheless, as machines' technological attributes and cultural boosters promoted their use, the numbers of horses on city streets and farm lanes diminished steadily.

Both the numbers and the economic value of horses began to weaken noticeably just before America's entry into World War I. In cities, the population of horses and mules had increased between 1890 and 1910 from two to more than three million, but by 1930, fewer than one million remained, largely replaced by other forms of motive power.[36] Manufacturing and public transportation led the way: in 1880, total inanimate horse power used in manufacturing surpassed that of animals; by 1907, animal-powered streetcar lines, stagecoaches, and omnibuses had almost ceased to exist in American cities.[37] The steepest decline in horse populations occurred after 1910 when trucks and automobiles became more commonly used for taxicabs, business deliveries, and hauling. Nationwide, horse values had peaked in 1910 and 1911; moreover, war demand (which began with European purchases of animals in 1914–15) seemed certain to drive prices up. However, the war only temporarily staved off the greatest losses and prices did not reach 1910 values during the war. By 1920, price per head had fallen 19 percent from 1910 levels.[38] The worst was yet to come; as it had in 1893, the overall economic situation and horse prices declined together. The postwar agricultural depression began in 1920, and by 1925 horse prices had dropped at least another 30 percent. Almost everyone connected with horse production (raising and selling), in the city and countryside both, felt the economic blow in the 1920s, if not before. Especially in rural centers of horse production, the decreased value of horses intensified the desperate circumstances of the nationwide agricultural depression.[39]

Once horses were no longer profitable to produce, many livestock traders and farmers began to slaughter excess animals, hoping to recover some money from the factory conversion of horse carcasses into glue, leather, and even dog food.[40] Census statisticians estimated that throughout the 1920s more than 200,000 horses were killed yearly on farms and in packing plants. In vain, BAI chief veterinarian Alonzo Melvin urged Americans to overcome their cultural taboo against eating horse meat, especially since the nation was enduring a meat shortage between about 1907 and 1915. By 1930, only a few pockets of working horses remained in service; most remaining horses served recreational roles or were kept because their owners valued their personal relationships with them. Combined with a tremendous decrease in horse production, the elimination of millions of "surplus" economically worthless horses cut the total American horse population by almost 40 percent between 1910 and 1930.[41]

Responding to Crisis

For veterinarians, this early twentieth-century collapse of the once-vibrant horse economy represented a threat to economic success and even to professional survival. Veterinary practitioners had naturally concentrated on animals whose monetary value encouraged their owners to seek professional medical care for them, and before World War I horses and mules brought twice the price of cattle and twenty times that of sheep or pigs.[42] Horses had also performed important business and transportation functions, and delays caused by their illnesses cost horse owners inconvenience as well as money; thus veterinary practitioners had been able to count on ministering to horses and mules that provided essential commercial functions. Everywhere, but especially in American cities, veterinary practice had depended heavily upon horses, the animals they recognized as "the most profitable veterinary patient."[43] The loss of equine patients not only affected the economic success of individual practitioners but also threatened to destroy the professional culture that veterinarians had so carefully built in the late nineteenth century.

During the years of World War I, veterinary education and (to some extent) practice still focused on the equine species and used traditional therapies. Veterinarian/writer James Herriot later recalled the "strange thrill in meeting with the age-old conditions whose names rang down almost from medieval times. Quittor, fistulous withers, poll evil, thrush, shoulder slip— vets had been wrestling with them for hundreds of years using very much the same drugs and procedures as myself." Horses and their ills created a continuity and identity for veterinarians—not always a welcome one, as leaders of the newly organized profession chafed against the old image of the horse doctor. Yet, under pressure from practitioners who comprised the majority of veterinarians, professional leaders had allowed the institutionalization of the traditional equine focus in the educational curricula and the state practice acts.[44] Given this intellectual and sociological dependence on equines and their work, the economic and cultural devaluing of horses promised to dramatically alter professional veterinary culture. This was especially true since most Americans continued to view veterinary medicine as synonymous with horses through the 1920s. Back in the 1890s—the early days of the transition to motor power—a *Turf, Field, and Farm* writer lamented that "with the diminishing use of the horse, the veterinarian is bound to find his field of future operations a circumscribed one indeed."[45]

The apparent prospects for veterinary education and practice deteriorated during and after World War I, when schools lost faculty and students, and practitioners' businesses were disrupted or abandoned. However, the depredations of the war did not compare with the effects of the transition away from horse power.[46] By the mid-1920s, veterinary educators occupied the unenviable position of trying to counter a public assumption of the profession's total demise. In 1925, a university colleague asked Cornell Veterinary College dean Veranus Moore how much longer he expected the veterinary college to remain in operation now that the horses were gone. Moore increasingly found himself addressing individuals and groups alike in an effort "to explain why veterinarians are still needed and why schools for their preparation and for research in animal diseases should be continued and liberally supported" in the wake of the gradual disappearance of horses from streets and roadways.[47] Public opinion presented a potential problem for veterinarians arguing for Bureau of Animal Industry appropriations, practitioners trying to make a living, and above all, for the veterinary schools and their deans and proprietors.

At the same time, decreasing veterinary school applications (especially after the war) alarmed educators and professional leaders. Although the problem of declining horse power loomed the largest, it was one of several factors that contributed to reduced enrollments at veterinary schools. Enrollments dropped after 1919, for example, when the War Department began requiring that potential Veterinary Corps officers have a high school degree for matriculation to veterinary school. Private veterinary schools, most of which had not required a high school degree for matriculation, suffered the most because they were dependent upon student tuition. They were also located in large cities and had trained their students to practice on horses.[48] With their sphere of usefulness shrinking, the numbers of applications to these schools dropped drastically after World War I. Citing difficult financial conditions, most closed their doors around the time of World War I, and none remained after 1927.[49]

Veterinary schools associated with universities also received fewer applicants and enrolled fewer students. Between 1914 and 1924, the total number of veterinary students in North American schools fell 75 percent. In 1921, ten of North America's sixteen veterinary schools graduated 15 or fewer students each; by 1926, the thirteen remaining schools in aggregate contributed only 130 new veterinarians to the profession. The magnitude of the decrease becomes apparent when this total is compared with the 136 students graduated in one year from just one school, the Ontario Veterinary College, 30 years earlier.[50] As R. M. Staley admitted ruefully in 1924, "judged by the attendance at our veterinary schools, the profession is not in danger of being over-crowded for some years to come."[51]

Even the most loyal source of applicants for veterinary schools—the families of veterinarians—proved reluctant to place faith in the future of their profession. As early as the beginning of the transition to motor power, some veterinary practitioners had been reluctant to send their sons (no women had yet graduated from a veterinary school) into a profession that might have no future. "Will it [veterinary medicine] prove an enduring profession in the sense of being a source of revenue to those practicing it? Here in the West, we hear a great deal about the passing of the horse," wrote one worried veterinarian to the *American Veterinary Review* in 1897. "I have a son who is now about the right age to enter college, and he is very anxious to begin his studies this fall. I should regret to have him spend three years at college and then find out that, like Othello, his occupation was gone."[52] By the 1920s, even this source of applicants had largely dried up; veterinarians' sons were not applying to veterinary schools. It is not clear if other, less desirable applicants—women and ethnic and racial minorities—were applying at this time to fill the vacuum. Certainly no evidence exists to show that veterinary schools recruited or admitted these applicants rather than go out of business.

For veterinarians already in the profession, the transition to motor power represented a professional crisis and a personal conversion. Practitioners anxiously scrutinized the horse markets and the continuing technological development of the automobile. Some panicked even before the century had turned. Certainly three discouraged Iowa practitioners believed the worst; in 1897, they left veterinary practice and returned to school for an M.D. degree, the better to support themselves.[53] In urban areas, practitioners watched uneasily as their most important patient population dwindled. By 1926, the dean of a Kansas veterinary school wrote that "mechanical motive power has made sufficient inroads upon the city veterinarian's practice as to make the horse virtually a negligible factor."[54] Horses had become so worthless that their owners no longer could afford much veterinary care for them. As a result, attrition in the ranks of urban equine practitioners contributed greatly to decreasing numbers of practicing veterinarians overall.

Despite their professional loyalties, even veterinary practitioners (especially rural ones) replaced their own horses with automobiles in droves after 1910. The veterinarian who left Philadelphia, Arthur Goldhaft, sold his horse and carriage and bought a Ford when he moved to the countryside to become a farm animal practitioner in 1916. The department of surgery and obstetrics at Cornell's veterinary college asked the university administration for an automobile in 1912 because "our competitors [local veterinary practitioners] use automobiles regularly in their practices." As one New York veterinarian later remembered, "in 1910 country practice in New York State was nearing the end of the horse and buggy age."[55] It was no coincidence that rural veterinarians purchased automobiles. With a more varied patient population, their clients would not view their abandonment of horse-drawn transportation with as much cynicism as would those of city practitioners engaged solely in horse practice. Automobiles were more practical for a rural practitioner's work, which was characterized by long drives in all conditions. The evident utility of automobiles only increased veterinarians' anxiety about the decreasing importance of horses, their own economic prospects, and the future of their profession. As one practitioner gloomily asserted in 1925, "I do not know whether [the veterinary profession] will live or die."[56]

Alarmed by this situation, some practitioners and educators began the process of determining how veterinarians could more actively control their professional destiny. The transition from horse to motor power taught veterinarians, as no other event had, that animal value functioned as a potent determinant of the procedures of everyday practice, research agendas, and even professional survival. In reaction to their circumstances, veterinarians sought to become more active mediators of the changing relations between Americans and their domestic animals. They entered debates about the position of the horse in American society, stressing the disadvantages of motorized transportation and finding niches that horse work could still occupy. Educators cultivated a close relationship between the Bureau of Animal Industry and the veterinary schools in order to secure veterinarians' place as public health workers. Finally, individual practitioners, encouraged by veterinary leaders, decided to begin emphasizing other species, along with the horse, in their practices. These efforts in reaction to the horse crisis redefined the role of veterinary medicine in American society for the remainder of the twentieth century.

Veterinarians' public efforts to support horse use, such as organizing urban workhorse parades, paralleled the activities of veterinary journal editors beginning in the late 1890s. Editorials sought to advance the cause of horse power by stressing the disadvantages of what they considered to be the horse's greatest competitor, the automobile. In 1897, the *American Veterinary Review* asserted that the horse "will outlive the present depression" (referring to a drop in horse prices connected to the national economic recession) and published stories about the mechanical unreliability of early automobiles. "There is no more danger of the displacement of the horse than there is of the extermination of man," declared the *Review*.[57] While the journal might be accused of naiveté and wishful thinking, its comments reflected the fact that the role of motor transport in American life remained uncertain before about 1910. Automobiles were expensive, rare, and mechanically frail, and gave the appearance of being a hobby, novelty, or fad. The *Review* gleefully reported on failed distance races, businesses that discarded broken-down motor trucks and bought more horses, and newspaper articles conciliatory to horse interests. The journal apparently did not suffer a lack of evidence for its claims about the difficulties of motor power and the reliability of horse power.[58]

Moreover, the veterinarians who edited the *Review* reminded their colleague-readers of the cultural value of horses, reflecting popular cultural concerns about the mechanization of American life. How could the attributes of automobiles compare with the "life, interest, and pleasure of man's best friend?" Surely no relationship with an inanimate object, however powerful, could reach the aesthetic and emotional heights of a "friendship." Indeed, national cultural identity required that the partnership between human and horse continue: the *Review*'s editors compared the replacement of horses by automobiles with that of the substitution of "incubators" for "American mothers." American culture, they asserted, would be the worse for either change.[59] This bizarre comparison echoed widespread critiques of the new technologies turning human society into a sort of machine, with individual humans losing identity as mere cogs within it or even being literally replaced by machines. This view proved quite threatening to many Americans brought up on nineteenth-century ideals of individual independence and distinctiveness in social and political relations. Although it would be too uncomplicated to call veterinary medicine a bastion of traditional liberalism, certainly its professional cohort of hard-nosed individual practitioners would have understood—and sympathized with—a warning about the perilous connection between technology and loss of individual humanity and autonomy.

As motorized vehicles became more common, veterinary commentators reframed the debate by insisting that employment existed for both motorized vehicles and horses, and that the acquisition of one need not entail the replacement of the other. Each mode of transportation had its particular niches. Horses remained the more economical, the most "elegant," and the least likely to produce "mental strain" in their drivers. Automobiles could provide pleasure, speed on a good road, and the ability to cover longer distances; motorized trucks could haul more weight. As interpreted by veterinary leaders, these advantages and disadvantages were transferred to American social stratifications. They predicted that small businessmen and professionals with geographically limited practices would retain horses; upper-class urban residents would also do so in order to "present the same smart appearance" that their "well-appointed broughams and horses of aristocratic bearing" had always created for them.[60] The potential uses of motorized vehicles, according to these calculations, would not erase all established uses of horses and would ensure some viable market niches for urban veterinary practitioners.

With these articles, the editors of the *American Veterinary Review* and other journals sought to enlist their readers as foot soldiers in the battle to retain roles for horses in the domestic animal economy—and to preserve the veterinary profession. "Talk horse everywhere," charged a *Veterinary Medicine* editorial, "and tell the world how much a horse can do." Journal articles provided information for practitioners who had to answer questions about the future of the horse economy and their profession on a daily basis.[61] Beginning in the late 1890s, their journals also encouraged practitioners to remain supportive of their profession during difficult times by performing their work to the highest standards and by sending family members and friends to veterinary schools. The father who had written to ask if he should encourage his son's interest in veterinary medicine in 1897 received a reply printed in the *American Veterinary Review*. The editors used this letter to express their confidence in the future of veterinary education and professional life, but hedged somewhat on the issue of the horse's continuing importance in American commerce. They addressed all of their colleagues when they assured the anxious father that he should "send his boy to the best veterinary school he may know, to thoroughly equip him for the practice of a profession whose lines are continually being enlarged." Yet they acknowledged that in some areas of work and commerce the horse might be "supplanted by mechanical appliances when they become perfected and cheaper." Nonetheless, the horse could be expected to remain the typical veterinary patient.[62] This staunch reassurance may not have comforted the father (who admitted that he had never seen an automobile and thus could not judge his adversary), but it clearly staked out the position of veterinary leaders at the beginning of the twentieth century. The horse, they asserted bravely, would remain valuable enough to continue as the cornerstone of veterinary education and practice.

Nonetheless, veterinary educators began quietly reforming their curricula to reflect new realities in the first decade of the new century. They had watched the "professional mystery" of laboratory science–based practices provide a new intellectual and educational program that had transformed their sister profession, human medicine.[63] They believed that "scientizing" veterinary medicine would help to attract better and more numerous applicants and support veterinary medicine's professional claim of expertise on animal health issues. With horses declining in numbers and value, the domestic animal economy's largest remaining populations of animals were those used for meat, milk, and other products, and their diseases threatened the wealth of the livestock economy and the health of human consumers. Conquering livestock diseases required the use of tools that were still being developed in laboratories, and reform-minded educators relied increasingly on the laboratory sciences for the proper training of veterinarians.

Veterinary educators settled on bacteriology, immunology, pathology, and physiology as the laboratory sciences most likely to help veterinarians solve animal disease problems and secure an identity as public health experts.[64] The veterinary schools at Iowa State, the University of Pennsylvania, Ohio State, and Cornell University already emphasized bacteriology in their curricula at the turn of the century.[65] The science of "immunity," by the mid-1890s, had helped to produce useful biological tools, including tuberculin, and the promise of other diagnostics and treatments. "The laws of action and reaction of the tissues to bacterial bodies," Cornell dean Veranus Moore wrote, would inform "a most intensely practical phase of professional work," providing a mechanism needed by veterinarians to become more efficacious as farm-animal practitioners and public health experts.[66] Comparative pathology, too, promised to become a cornerstone of attempts to understand diseases transmitted between food-producing animals and human consumers (zoonoses). Veterinarian L. H. Pammel wrote in 1906 that a thorough scientific training in pathology and physiology would ensure an important place in every community for veterinarians displaced by the horse crisis. Accordingly, the veterinary school at Iowa State spent almost as many hours on pathology as on the equine-oriented theory and practice of medicines in its 1905 curriculum.[67] The school's animal physiology laboratory, built in 1912, "represent[ed] the last word in this branch of veterinary science"—a branch expected to grow as practitioners shifted their focus from horses to farm animals.[68]

The reduced emphasis on equine medicine and anatomy, the continuing addition of the laboratory sciences and animal husbandry subjects, and the expansion of courses devoted to food inspection and other public health concerns all exemplified changes in veterinary education that reflected the declining value of horses.[69] Of course, not all schools could afford the laboratories and other facilities necessary for these reforms; nor did all veterinary educators support them. Many of the urban proprietary schools lacked both funds and interest in building new laboratories and refocusing on food-producing animals, and this was one of the reasons that all of these institutions had closed their doors by the end of the 1920s. The vast majority of surviving veterinary schools in 1930 were independent units within land-grant universities, near other scientific departments and large populations of farm animals.[70] Many practitioners and potential students disappointed by these changes in veterinary education interpreted the reforms as devices that would intimidate and exclude the farmers' and tradesmen's sons who had traditionally been drawn to veterinary medicine. Opponents to reform were insulted by veterinary leaders' equation of potential professional success with intellectual ability and social status, and they responded by criticizing stringent matriculation requirements and the emphasis on laboratory-based education.[71] Overcoming their opposition would require not only the desperate circumstances of the horse crisis but also the intervention of the BAI.

BAI Control

The BAI assumed a much greater role in shaping the veterinary profession as a result of the crisis of the transition to motor power. Since its inception, the BAI and the veterinary profession had always been closely related, but as long as the majority of veterinary practitioners (urban, equine oriented) had goals divergent from those of BAI veterinarians (researchers of primarily food-animal diseases), the BAI's influence within the profession was limited. The potential for BAI control of the profession had always been present, however. Prominent veterinarians who functioned as leaders of the profession included educators, officers of the BAI and the American Veterinary Medical Association and state associations, and state and experiment station veterinarians. Many of these professional leaders had been BAI personnel at some point in their careers, and they embodied the close ties between the BAI and the veterinary profession.[72] Veranus Moore, before joining Cornell's faculty, had headed the BAI's pathology laboratory; BAI chief officers Daniel Salmon, Alonzo D. Melvin, and John R. Mohler served also as officers of the AVMA. These men and others occupied key positions in both organizations and they had the ability to influence the direction of both. They focused on reconfiguring the veterinary profession by engaging the BAI in their drive to reform veterinary education.

Beginning in 1907–8, the source of the BAI's power to contribute to educational reform lay with its promise of employment in a tightening animal health marketplace. During the transition from horse power, veterinary medicine needed an assurance of steady employment for its members, or at least a recognizable market niche with which to attract professional aspirants. It found both in the BAI. Over time, the BAI was the single largest employer of veterinarians in the United States, and during the post–World War I period of low market demand it provided crucial stability for the profession. In 1908, the BAI employed more than 800 full-time veterinarians, or approximately 7 percent of the total number in the United States. Its veterinary force continued to grow; in 1921 it employed 1,550 full-time veterinarians, or about 10 percent of the entire profession (a proportion that continued into the 1930s). These numbers did not include the large population of part-time workers (often uncounted by the BAI)—private practitioners doing specific contract work in their local areas. Whether doing full- or part-time work, veterinarians depended on the steady income that BAI jobs brought them. Veterinary leaders pointed to BAI jobs as examples of promising careers for young veterinarians—a strategy designed to address the popular perception that the profession had no future outside the equine sphere. Practitioners interested in urban horse practice had previously ignored the BAI's requirements (implemented in the 1890s) for a civil service examination and graduation from a bureau-approved veterinary school. In the dwindling horse economy, however, both of these requirements took on new importance, as demonstrated by the many veterinary school applicants who wrote to the BAI for preapproval of their educational plans in the 1910s and '20s.[73]

BAI employee veterinarians worked as meat inspectors and scientists and depended heavily upon their knowledge of pathology and bacteriology. Thus BAI-accredited schools were required to emphasize these subjects in their curricula, as Iowa State, Cornell, and the University of Pennsylvania had done. By placing veterinary education solidly on the base of these and other laboratory sciences, veterinary professional leaders provided the BAI with appropriately trained employees. In turn, satisfying the BAI's needs for trained personnel helped the veterinary profession survive the crisis of declining horse values and populations. Not only did the BAI provide steady employment, it was also the major mediator of the profession's relationship with the state. Questions of federal funding for animal disease eradication projects and legislation affecting the profession often involved the BAI, making it an important early lobbyist on behalf of veterinarians.[74]

By 1910, the BAI had used its market position to win surprising powers of oversight in the reform of veterinary school curricula and the practical aspects of veterinary education. Anticipating the much more well-known Flexner Report, Secretary of Agriculture James A. Wilson (the cabinet-level head of the BAI) appointed a committee to inspect and report on all U.S. veterinary schools in 1908.[75] The committee, which included representatives of the AVMA, BAI, and Association of Veterinary Faculties, rated all schools and categorized them according to their acceptability to the BAI as trainers of its employees. BAI leaders and educational reformers agreed on the basis and outline of an appropriate veterinary education, and they created a new covenant. The BAI could, under the aegis of employment training and marketplace control, carry out reformers' educational program, enforce it through the civil service examination requirements for veterinarians wanting to become employees, and provide a stable source of veterinary employment. Forgoing the powerless AVMA and warring state boards of examiners, veterinary professional leaders cooperated with the BAI to achieve those desired goals, and the BAI was in return guaranteed a trained labor supply necessary to its continued growth (which was especially important after the Meat Inspection Act of 1906 called for more veterinary inspectors).[76] If they were rated "acceptable," a veterinary school's graduates were eligible to take the civil service examination for BAI veterinary inspection jobs. Eight schools did not make the list, thereby effectively excluding their graduates from ever applying for a BAI job.[77] Rejected schools faced the loss of students wary of limited career opportunities, and the wrath of graduates who suddenly found themselves ineligible for employment with the BAI.[78]

The committee may have been grudgingly excused for judging schools' overall fitness to train BAI employees, but many veterinarians believed it overstepped its bounds with regulations governing students, college administrations and faculties, and curricula and courses of study. While the list of approved schools generated concern, the committee's other recommendations, once transmuted into official regulations by Secretary Wilson, caused an uproar.[79] In order to stay on the "acceptable" list, veterinary schools had to comply with detailed regulations, including clearing decisions about classes, textbooks, and the admission of students with the BAI. The BAI's personnel department kept a file that contained a card for each student admitted to a veterinary college after 1908, and veterinary colleges reported the names and characteristics of students enrolled every year.[80] In this way, the BAI tracked the matriculation qualifications, age, and professional movements of every student in the United States (whether they graduated or not). Regulations also stipulated faculty qualifications, including restricting the number of professors who had been trained at the same school. Schools were prohibited from holding more than one graduation per year or matriculation examinations on any nonapproved date. The course of instruction had to take place over 3 years with not less than 6½ months per year, or 3,000 hours of instruction.[81] Such was the BAI's determination for control that it discouraged any initiatives on the part of the schools to alter their requirements, even to make them more stringent.[82]

In return for its control, which guaranteed the education of employees at no cost to the USDA, the BAI stipulated that only applicants who had received a diploma from a veterinary college could take the civil service examination for the position of veterinary inspector, thus keeping its covenant with the profession to employ only veterinarians as food inspectors. For better or for worse, the BAI's extensive power to regulate veterinary schools continued for 20 years after the original regulations went into effect in 1909 because it controlled a significant part of the market for veterinary services. The BAI had taken advantage of veterinarians' reduced opportunities in the private market as horses declined in value. The BAI-enforced reform of veterinary education, which set the profession on a new intellectual and sociological path for the rest of the twentieth century, originated in the desperate circumstances of the horse-power transition.[83]

Entering the Arena of Public Health

While the BAI was the single most important savior of veterinarians during the transition to motor power, private practice (albeit in a different form) remained significant as well. Urged by the veterinary journals to explore every professional opportunity, many equine practitioners acted on a new strategy: they sought to make patients of species of animals other than horses. They were encouraged by Cornell professor Pierre A. Fish, who reminded disconsolate veterinarians in 1922 that opportunities still existed outside of the equine sphere: "In the commercial world there have been instances where changed conditions have rendered it necessary for manufacturing plants to adapt their machinery and equipment to a totally different kind of production if they were to continue their existence. There has been a similar necessity for veterinarians, on account of the diminishing horse practice, to adapt themselves to the increasing demands for their services in relation to the other domesticated animals."[84] Urban practitioners did not disappear from cities despite decreasing horse populations and value; Atlanta's city directory listed 5 veterinarians in 1899 and 6 in 1935, and Philadelphia counted 86 in 1900 and 71 in 1925.[85] Veterinarians who remained in, or migrated to, cities during this period instead diversified their practices to include other urban animals—dairy cows, stockyard inhabitants, and companion animals. Companion animals, or pets, could "constitute a considerable portion of the practice," wrote Pierre Fish, "and such work is often quite lucrative." Some successful urban veterinarians weathered the crisis of dwindling horse practices by slowly transforming their old facilities designed for horses into diversified pet hospitals. Others built new animal hospitals designed specifically for companion animals.[86] The numbers of pet hospitals built or renovated by private veterinarians swelled in the 1920s, attracting interest among city dwellers with pets. Although companion animals still represented the smallest patient population and source of income by far, Fish could report by 1922 that "numerous veterinarians derive their income from this source."[87] Work in the sanitary sciences, including milk and meat inspection, represented another possibility for employment. Urban veterinarians, who still made up the bulk of the organized profession as late as World War I, continued to care for the remaining city horses while actively developing new patient populations in order to survive economically.[88]

Rural veterinarians, whose practices had always been more diversified, began to shift their focus to food-producing animals, such as dairy and beef cattle, sheep, swine, and poultry. Even before World War I, professional leaders had seized on agriculture, and the animals owned by agriculturalists, as the next promising market for veterinary services. This choice made sense for several reasons. First, agriculturalists owned the most economically valuable animals: draft horses on farms represented the only remaining repository of horses in the country, and authorities asserted that livestock breeders were willing to "guard more cautiously" the health of their most valuable cattle, sheep, and hogs as well.[89] Second, veterinarians who hailed from rural areas found the idea of returning to country life acceptable. Finally, veterinary leaders were eventually able to articulate a vision of their profession's worth that was based on a wider role for veterinarians as guardians of the nation's health as well as its wealth. For former horse practitioners, a new focus on food-producing animals could provide financial security.

Unfortunately, this strategy backfired somewhat for private practitioners when an economic depression hit American agriculture in the wake of World War I.[90] Before the disruptions of the war, farmers had enjoyed a hypothetical economic "golden age" in the 1910s. Crops brought in returns greater than the farmer's cost of materials and labor. With their disposable income, rural people purchased everything from automobiles to veterinary services just before and during World War I.[91] Half of the nation's population still lived in rural areas, so the rural consumer market was theoretically quite large. Prosperity did not last long, however. The war disrupted traditional patterns of farm production as farmers were urged to feed armies and populations around the world. The resultant overproduction and neglectful federal policies collapsed farm commodity prices in 1919–20. Livestock prices also declined, and the veterinarians who had had rural practices before the war often returned to find their businesses disrupted. "Any reaction which depresses general prosperity," acknowledged Pierre Fish, "depresses the veterinarian as well."[92]

Nonetheless, veterinary professional leaders continued to support the idea that "the prosperity of the veterinarian" should be "linked with the prosperity of the farmer."[93] The depression could not last long, veterinarians reassured themselves; it was merely a part of the temporary economic chaos, a natural sequel to the war. "In the long run," asserted S. J. Schilling, "we have no need to fear for the life and plane of existence of the veterinary profession" dependent upon agricultural animals.[94] Veterinary leaders may have been uneasy at charting their profession's course as practitioners' warnings about desperate rural economic conditions reached them, but they had little alternative. Horse practice was diminishing and companion animal practice had not yet developed to the point that large numbers of practitioners could depend upon it.[95] By 1925, despite the agricultural depression, the veterinary profession's primary sites of production and practice, both physically and intellectually, had shifted to rural America and farm animals. Veterinary leaders helped to accomplish this reorientation by transforming the curricula of veterinary schools to emphasize the health and diseases of food-producing animals.[96]

Time-honored relationships between humans and horses had been redefined, forcing veterinarians who had depended on those relationships to take an active role in reshaping their practices, schools, and professional identity. Many found the experience personally traumatic. The survival of the veterinary sciences and the veterinary profession appeared, to some, tenuous at best. However, the horse crisis awakened in veterinary professional leaders a sense of responsibility for defining their discipline's identity amid new social realities. Fortuitously, the laboratory sciences linked veterinarians solidly to crucial public health issues from the late 1890s through the 1920s during the time that horse populations were declining. The crisis represented an opportunity for American veterinary leaders: they could more easily develop the profession of veterinary medicine as an applied science with a public health mission, given the weakened resistance of horse-oriented practitioners. Veterinary practice changed because the value of veterinarians' patients had changed; and in turn, more veterinarians understood for the first time how they might gain professional authority by helping to shape the domestic animal economy in the coming decades of the twentieth century.

In choosing to focus on food-producing animals and their ills, veterinary leaders gained a great deal (despite the difficulties of the postwar depression). First, using the laboratory sciences to study the diseases of food-producing animals fit into an already existing tradition established by institutions such as the BAI and experiment stations.[97] This strategy ensured intellectual and institutional connections between veterinarians and agricultural scientists. Second, veterinarians gained nationwide legitimacy as experts who could address the economic imperatives of food-animal production. This legitimacy brought with it a new practical focus on preventive medicine and herd health. Finally, and perhaps most crucially, a concentration on food-producing animals potentially provided veterinary scientists and practicing veterinarians with authoritative roles as public health regulators. Veterinarians argued that they were the experts best trained to judge the safety of Americans' food supply, and they would play key roles in creating and implementing public health regulations associated with meat, milk, and other animal products. Thus, the sphere and social mandate of veterinarians expanded greatly: besides being accountable for livestock owners' economic interests, veterinarians also seized some responsibility for human health—a moral obligation widely understood to be of the highest order.

CHAPTER THREE

The Value of Animal
Health for Human Health

"Are you willing to do what you can to protect the babies of your city?" health officer George W. Goler asked Americans at the beginning of the twentieth century. "Clean milk," he added, "is a necessity for babies; thousands of children under the age of five are dependent for their very lives upon milk." Directly linked by sanitarians and laboratory scientists to appalling rates of infant and child mortality, the quality of the commercial milk supply became invested with life-and-death importance as physicians and concerned citizens scrutinized the procedures used by livestock owners in the production of milk, meat, and other foods. Although disease could be caused by contaminants introduced into foods through careless handling, many of the most serious food-borne diseases were thought to have originated in the live animal. This demonstrable understanding of the organic relationship between the health of animal bodies and that of human bodies stimulated widespread public interest in the conditions under which animals lived, produced, and died. Food animals became valuable for their impact upon human health, and they became worthy of professional expert attention based on this value.[1]

Veterinarians had stepped forward to offer their expert attention to the problems of food-animal health even before World War I. However, the livestock producers who were their clients did not always welcome the scrutiny and restrictive laws that the veterinarian represented. "We know just as well as a veterinarian whether our cattle are sick or not," argued a New Jersey dairyman. "I do not want to see the New Jersey cattle in the hands of the veterinaries."[2] When acting as regulators, veterinarians could threaten livestock producers' profits by preventing them from selling their animals or animal products. The accompanying implication of unhealthy animals and poor husbandry offended producers. Finally, producers resented the power of the state and its agents (including veterinarians) to regulate their practices. Veterinarians interested in enforcing public health regulations thus risked alienating livestock producers—the very clients whom they hoped to cultivate as horse practice waned.

Between the 1890s and 1930s, veterinarians had to learn to be expert mediators of the value of animal health in American society, encouraging the cooperation of livestock producers and public health activists. Conflicts between the two groups revolved around the regulation of the food supply. Municipalities, states, and the federal government all developed programs that sought to define a "healthy" animal and "clean" foods of animal origin, outline the steps necessary to produce such animals and foods, and provide for implementation and enforcement of production rules. For regulations to succeed, urban consumers had to understand that they were bound to rural producers, and livestock owners and industries needed to agree (however reluctantly) to increased state regulation as a measure to restore consumer confidence. Both groups engaged scientific experts, but on their own terms, to help define the parameters of regulatory success.[3]

Veterinarians engaged as food supply experts often found themselves at odds with concerned consumers as well as with livestock producers. As with other early twentieth-century social reforms, the most well-known food regulation activists were upper- and middle-class urban women. These women associated animal welfare with healthy food, created definitions of "natural" and "wholesome," and extended standards of class and gender-based behavior to other participants in the debate over healthy foods of animal origin. The masculine barnyard culture that many veterinary practitioners were accustomed to clashed with the ethos of female activists who had their own ideas about proper animal treatment, definitions of "healthy" animals, and procedures in milking facilities and slaughterhouses. Despite these differences, veterinarians interested in public health work espoused the activists' views and sought compromises with livestock owners. This mediatory role promised to broaden the intellectual and professional scope of veterinary medicine beyond "horse doctoring." It also allowed veterinarians to appropriate scientific tools and to join their medical brethren in shaping public policy at local, state, and national levels.

Regulating the Municipal Milk Supply

The regulation of the milk supply provided an important opportunity for veterinarians to join physicians and other officials in public health reform. Fighting outbreaks of infectious disease traced to cities' milk supplies called for strict regulatory measures. "It means," asserted a Buffalo, New York, physician, "a system feasible for daily surveillance and effective in the protection against frequent infection of the milk-consuming denizens." The plan for such a system usually included the inspection of dairies, the registration of milk dealers and sellers, and the use of laboratory tests to monitor milk quality. However, the health officials of many municipalities at the turn of the century bemoaned the absence of such controls. In major cities such as Omaha, St. Louis, and Detroit, health officials reporting on the milk supply in 1903 put high priority on establishing mandatory veterinary inspection of all dairies supplying the city—including those in rural areas.[4]

Inspection of rural producers of milk to be consumed by city dwellers was a tangible sign of the early twentieth-century city's reach over the surrounding area that supplied its provisions. Rural and agricultural hinterlands sometimes extended for hundreds of miles, especially in the case of western cities like Chicago and Denver; urban and rural places were not necessarily distinct entities with clear boundaries.[5] Animals and animal products crossed freely between city and countryside, and urbanites came to recognize that geographic boundaries were surprisingly permeable not only to animal products but also to the impurities and diseases they carried with them. Inhabitants of the hinterland, while grudgingly acknowledging the city markets' influence over their production choices, nonetheless fought to retain some autonomy.

Most cities began regulating the quality of milk produced inside and outside of their boundaries during the decade of the 1890s. As early as 1895, St. Paul, Minnesota, authorized the city council to provide by ordinance for the inspection of milk, dairies, and dairy herds, and to issue licenses for milk sales within the city. Colorado Springs and Denver did not decide upon milk regulation until 1897 and 1899, respectively; both prohibited the sale of "impure" and "adulterated" milk within city boundaries, regardless of its place of origin. Municipal ordinances sometimes reinforced state laws regulating the milk supply. Such was the case with Philadelphia, which approved a ten-section ordinance in September 1890 prohibiting the sale within the city of impure or adulterated milk; milk from cows living in a "crowded or unhealthy" condition, exposed to tuberculosis, or fed "swill" (by-products of brewing and other manufacturing).[6]

These municipal ordinances regulating the quality of the milk sold in the city applied to both urban and rural commercial dairies, but not to private or family cows. Many cities at the turn of the century counted numerous dairy cows among their inhabitants. San Francisco, for example, boasted one cow living in the city for every five human inhabitants. Urban families were generally allowed to retain up to two uninspected cows as long as they did not openly sell the milk. In Minneapolis, these "family" animals made up 48 percent of the cows living in the city, so a significant number of residents consumed an unregulated milk supply. Larger milk producers within city limits protested that the inspection exemption hurt their business by creating a monopoly in favor of persons who kept one or two cows who could use or even quietly sell milk without inspection or licensure. Tightening regulations to prohibit the selling of milk from "two-cow" operations resulted, but this source of unregulated milk did not significantly decline until municipalities began to regulate the keeping of animals within city limits.[7]

Cities adopted two types of milk supply regulations in the 1890s. First, the quality of the milk itself was subject to inspection. Most municipal ordinances prohibited the sale of impure or adulterated milk (the definitions of which were open to interpretation). Impurities usually meant particles, such as mud or manure, that had inadvertently fallen into the milk. Impure milk also came from cows who had been fed poorly, especially with the by-products of the distilling or brewing industry. Adulterants were assumed to have been added deliberately, and included preservatives such as formalin to keep milk looking and smelling fresh. Milk standards also often specified the minimum percentage of butterfat and solids allowable. Before 1910, most prosecutions under municipal regulations concerned sellers who had misled consumers with watered-down or artificially preserved milk. Such regulation was designed to keep the consumer from being cheated and to preserve the taste and aesthetic qualities of the milk, with little attention to health concerns. Penalties for milk dealers (and city producers) found guilty of watering or skimming milk were usually the loss of a license and perhaps a fine.[8]

The second type of regulation in the 1890s dealt with the conditions under which milk was produced and handled before reaching the con-

THE UNREGULATED MILK SUPPLY: A BACKYARD COW IN TOWN, CIRCA 1910. The "family cow" provided many urban residents with milk. These cows usually escaped inspection and worried health officials interested in controlling the urban milk supply. *Courtesy of Coll. BHS 207-1-18, "802 Arapahoe Avenue," Carnegie Branch Library for Local History, Boulder Historical Society Collection, Boulder, Colorado.*

sumer's kitchen. Requirements commonly included keeping the cows healthy and their stables and udders clean; discarding the milk from cows with newborn calves; and barring persons ill with contagious diseases from handling the cows or the milk. Milk bottles and other vessels, and the areas in which milk was handled, had to be kept clean. Such sanitary regulations reflected a general understanding of the portability of disease arising from sick cows or handlers, through the medium of milk, but their primary emphasis was on the need to keep unclean or impure milk from being unfairly sold rather than invoking the specter of tuberculosis or other diseases. Veterinarians and physicians decried the lack of concern for public health that these measures implied. "The public inspection of milk in the United States," charged the pasteurization advocate Nathan Straus in 1897, "is thus directed mainly to the prevention of fraud" rather than to the guarantee of healthfulness.[9]

The bacteriological understanding of tuberculosis and other infectious diseases began to alter the atmosphere of municipal regulation in the first decade of the twentieth century.[10] First, the parameters defining milk's fitness for consumption—the definition of "healthy" milk—changed. The health or disease, not just the cleanliness, of the cow and her handlers became important to consumers wealthy enough to be more worried about tuberculosis and typhoid than the price of milk. This concern was reflected in the debate over the standard laboratory tests utilized to judge the fitness of milk for city consumption. In the last decade of the nineteenth century, inspectors had most commonly submitted milk samples to chemists who performed the Babcock and lactometer tests (measuring parameters such as the percentage of butterfat and adulterants), both of which reflected a concern that the milk not be watered down or falsely preserved.

Dissatisfied with such crude investigations, public health advocates not only called for detection of poisonous adulterants, but increasingly focused on bacteriological testing as a measure of milk's wholesomeness. Municipal utilization of bacteriological testing progressed very slowly, however. Of thirty-eight large cities polled in 1903, only ten had conducted more than fifty bacteriological tests on milk samples in the previous year. Los Angeles, Washington, D.C., and Cincinnati, among others, had conducted none at all. It would be tempting to attribute the lack of bacteriological analysis solely to milk inspectors lacking scientific training, but Washington and Cincinnati's responding health officers were both physicians, and Washington also listed two full-time veterinarians on its payroll. More likely, a lack of interest within the health department, a nascent but insufficient consumer demand, and resistance from milk producers and dealers combined to inhibit the testing.[11]

Bacteriological testing was an important early standard, however, for advocates of certified milk. Amid physicians' worries that the nutritional qualities of "boiled" milk would be diminished, pasteurization was viewed with suspicion and not widely employed before 1910. The certified milk movement experienced its heyday between about 1900 and 1918. Certified milk, so accredited by special milk commissions and municipal health officers, had to be produced under stringent conditions of cleanliness, bottled in sterile containers, and maintained under strict bacteriological testing. The highest allowable bacterial counts per cubic centimeter of milk varied from place to place, but most ranged between 100,000 and 500,000.

Therefore, production of certified milk depended upon the existence of two institutions: a local bacteriological laboratory and farms able to meet the stringent sanitary conditions. The milk also cost more, and the certified milk movement's success depended in part on consumers who were willing to pay up to twice as much money per quart. Those who could afford the milk and who valued cleanliness and the health of food-producing animals did not complain.[12]

Certified milk also required a willing producer: one who expected to maintain the necessary profit margin, but also one tolerant of scrutiny. "I confidently believe that if the farmers were to receive five or six cents per quart, and were told how to produce clean milk, many of them would do it," asserted a Boston health official. However, the rural dairy producers surrounding the large cities found that production of certified milk did not always increase profits and came with the additional price of losing authority in one's own milking parlor. Bacteriological knowledge changed the regulatory climate of the dairy itself, to the alarm of dairymen. Rural dairymen selling milk in many cities faced an examination of the cleanliness of their stables and cows. Others desiring to sell milk certified by the local medical milk commission had to follow physicians' detailed instructions stipulating whitewashing frequency, employee apparel, and bovine management.[13]

The greatest threat to a dairyman's autonomy and economic bottom line lay with the identification and condemnation of tubercular cows. Such cows were identified by a certified veterinarian who administered the tuberculin test. Between 1895 and 1917, most states and larger municipalities passed laws requiring tuberculin testing of milk cows. Although scientists generally agreed that the most dangerous cows were those with tubercles in the udder, any animal that tested positive was usually condemned to slaughter regardless of clinical signs. Depending on his state of residence, the unfortunate dairy producer often received only a part of the animal's economic worth in recompense. Unhappy rural and urban dairymen rallied their resistance to all milk regulations around the issue of veterinary inspection and tuberculin testing. Turn-of-the-century dairy producers singled out veterinarians as a target of their wrath, accusing the profession of collecting inspectors' fees while condemning producers to economic loss.[14]

As a part of their antiregulatory campaign, dairymen also questioned veterinarians' scientific knowledge. The use of bacteriological agents for diagnostic testing was still in its infancy. Many farmers had never seen the application of tuberculin or the gross or microscopic lesions characteristic of tuberculosis. Some of the cows condemned by the tuberculin test looked perfectly healthy, creating suspicions that the test was a flawed or arbitrary measure of animal health. Most veterinary inspectors were hired on a part-time, contract basis while maintaining their full-time private practices. Those certified to administer tuberculin tests, a skill acquired at veterinary schools, were generally younger graduates trained in bacteriology. Their rural clientele's apparent suspicion of newfangled tests and wet-behind-the-ears practitioners probably accounted for much of the opposition to tuberculin testing. However, producers' criticisms also reflected the fact that veterinarians had to maintain carefully standardized procedures. Individual cows' reactions to tuberculin—fever and swelling at the injection site—could sometimes be open to differing interpretations. "It takes some time, patience, and blunders," admitted D. H. Udall, a Vermont veterinarian and inspector, "to get things running smoothly and adjust the scientific treatment to field conditions"—a situation that farmers did not tolerate without complaint.[15]

Dairy producers losing money on condemned animals resorted to the courts in an attempt to challenge mandatory tuberculin testing. In 1909, the city of Milwaukee adopted an ordinance requiring that dairymen selling milk in the city file in the health commissioner's office a certificate from "a duly licensed veterinary surgeon" stating that their cows were free of tuberculosis and other contagious diseases. Every year, each cow had to be numbered, described, tested, and pronounced healthy. Under state law, cows positive for tuberculosis were condemned; under the new municipal ordinance, milk from untested, unregistered herds would be confiscated and destroyed with no apparent recompense to the farmer. Rural dairymen argued that urban producers escaped the penalties of the ordinance, since their cows only required infrequent inspection by the "health officer," not the dreaded tuberculin test administered by the veterinarian. Outraged rural milk producers found a champion in "John Quincy Adams," a dairy farmer who carried his suit against the Milwaukee ordinance all the way to the U.S. Supreme Court. "Adams" alleged, among other things, that the city health commission had no right to extend its jurisdiction to his milk and moreover that the veterinary test to be applied to his cows—tuberculin—he believed to be "wholly unreliable, untrustworthy, and entirely worthless so far as being a guide or protection to the public."[16]

However, the Court found in 1913 that "local belief"—in this case, the dairymen's assertion of the tuberculin test's invalidity—could not over-throw a local regulation "found to be necessary for the protection of the public health." Moreover, the Court asserted that municipalities had the right to insist on veterinary inspection of cows and test milk that issued from rural dairies outside of their normal jurisdiction. In matters of public health, the Court warned, "there is no discrimination" between "milk from cows outside and milk from cows within the city." Even though city health inspectors could not themselves inspect rural cattle, they could send their field experts, veterinarians, to evaluate the cows as a measure of the milk's healthfulness.[17]

Within the regulatory framework, veterinarians controlled tuberculin testing, but they were not able to expand their monopoly much beyond this diagnostic function. Most cities relied on lay milk inspectors or assistants who sporadically peered into cow stables and noted the condition of the animals and workers. A Rochester health official suggested that field milk inspectors be women, preferably trained nurses, who could travel around inspecting dairies and report to municipal health authorities. Physicians also functioned as dairy inspectors. To the chagrin of veterinarians, some physicians felt justified in providing expert opinions on milk cow breeding and care and considered themselves qualified to examine animals for disease. As late as 1922, Louis Klein, a faculty member at the University of Pennsylvania School of Veterinary Medicine, lamented that many of his well-qualified graduate veterinarians had difficulty competing with (politically appointed) lay people for jobs as general milk inspectors.[18] Nonetheless, veterinarians were still the only ones authorized to perform the tuberculin test and condemn tubercular cows by law. Especially after the Bureau of Animal Industry commenced a federal-state campaign to eradicate bovine tuberculosis in 1917, veterinarians became indispensable as field experts who judged animal health in thousands of localities around the country. Through the 1940s, generous congressional appropriations for tuberculin testing (and compensation for condemned animals) resulted from heavy lobbying from both consumer and producer groups and guaranteed an important role for veterinarians.[19]

Producers' associations began to support legislation mandating tuberculin testing only after being assured of appraisal and economic compensation for condemned cattle. Once regulations were in place for compensation under the federal tuberculosis eradication program, the National Grange quickly "urged" producers to cull tuberculin-reacting cows from their herds. Major agricultural journals and breeders' associations agreed that testing would ultimately ensure both the physical health of animals and the economic health of the milk industry. However, the greatest influence on producers cooperating with the authorities arose from slowly mounting local pressure and tension within neighborhood relationships. The shift away from producers' earlier resistance occurred only after they became resigned to regulation and began to police their own local districts. Pennsylvania farmers surveyed in 1900 bitterly recalled that tubercular cows who later died or infected their other animals had been purchased from a neighbor, a well-known local breeder, or at a local public sale. "There are some of the people [who] think their cattle have it and [sell] them, then others must suffer the penalty," complained one dairy producer. In response to such dangers, producers of all sizes banded together over the next 20 years to form area testing sites, often a township or county, that could become accredited as tuberculosis free. Milk from accredited districts satisfied city ordinances, and the animals brought higher prices and were not restricted from shows and fairs (a special incentive for purebred breeders, who often led local testing efforts).[20] By World War I, dairy producers' protests had quieted around much of the nation as they acceded to regulations demanded by milk consumers.[21]

In common with other social issues, popular support for milk regulation gained tremendous momentum around 1900 (as historians have shown). By 1906, public health official C. Hampson Jones of Baltimore asserted that "the real control of the milk trade is largely in the hands of the public." Jones referred primarily to women and their organizations, who led the way in exposing many practices of the animal industries that were abhorrent to consumers. Interest in public health focused on concerns central to most women's lives, including the healthfulness of the family's home. Driven by well-publicized mortality rates among young children, mothers joined forces with physicians and nurses to form voluntary milk commissions, erect infant feeding stations, and encourage the production of certified milk. Large cities such as Rochester and Baltimore supported extensive educational campaigns that included lectures and exhibitions attended by producers and consumers alike.[22]

The call for clean milk for all children, regardless of socioeconomic status, also caught the imagination of settlement house workers and other social reformers. Those who worked to transform municipalities—healthcare professionals, social reformers, and citizens' coalitions—found in milk and meat regulation a cause that one reformer asserted would "strengthen the co-operative relations in the group of people who were responsible for the [urban reform] movement."[23] Historian Suellen Hoy has described reformers' interest in milk (and meat) as part of a larger municipal housekeeping program, championed by women who equated cleanliness with social order. Activists scrutinized and demanded changes in the whole process of animal production, food handling, sale, and consumption of animal products. To them, these components were inseparable and all demanded reformers' attention. They also explicitly connected the welfare of food-producing animals to the healthfulness of milk and meat. Along with his advocacy of pasteurization, Nathan Straus repeatedly stressed the connection between unhealthy cows and unhealthy milk. Physician and reformer Sarah D. Belcher decried the conditions under which dairy cows were commonly kept—locked up in stanchions or stalls with little or no access to a pasture, lying in manure, and eating unwholesome feed. The "welfare" of an animal affected "the quality of her product," Belcher asserted. In the opinion of Belcher and many other reformers, regulation of the milk supply had to include attention to the living conditions of cows as well as to the number of bacteria in the milk, and municipal regulations from New York to Denver to San Francisco included stipulations about both.[24]

Veterinarians contributed to reform efforts by creating a concept of public health regulation that involved the ideas and practices of both livestock producers and milk and meat consumers. The value of animal health to producers was an economic one, while for consumers it was a vision of "pure" food and healthy children. As Ilyse Barkan has demonstrated, successful long-term regulation could not be created without both parties finding something useful in it for themselves. Nor could regulation survive challenges without social institutions—in this case, science and Progressive Era reform—that provided authority to support it. Veterinary leaders drew on these sources of authority and their responsibilities to both producers and consumers when making health policy recommendations. This strategy drew criticism from early twentieth-century physicians (and medical historians decades later), who accused veterinarians of protecting livestock interests at the expense of progress in public health. With pressures from both sides, veterinarians occupied a uniquely uncomfortable position. Indeed, their situation would have been much more straightforward had they categorically advocated the livestock industry's views, but they did not. Veterinarians' role in the creation and implementation of health regulations was far more multifaceted than existing accounts imply.[25]

Bovine Tuberculosis and Meat Inspection

Three studies of reform at the municipal, state, and federal levels reveal the complex negotiations among veterinarians, livestock producers, and consumer advocates in securing regulation. The milk supply was not the only focus; many concerned citizens' groups concentrated on obtaining regulation of meat as well. "The great wave of municipal reform that has been sweeping over the breadth and length of this land," asserted veterinarian A. S. Wheeler in 1897, "will do much to bring meat inspection into popular notice." Invoking their right to ensure the healthfulness of their family's food supply, women reformers had ventured before the turn of the century into gruesome slaughterhouses to conduct "citizen's inspections." Few municipal officials (almost all men) were willing to accompany them, although at least one credible male witness appeared to be necessary to lend an air of decency and authority to an otherwise abhorrent task. In 1897, Philadelphia veterinarians Leonard Pearson and W. Horace Hoskins accompanied physicians, health officials, and members of the Women's Health Protective Association (WHPA) on an unannounced visit to a few of Philadelphia's 104 slaughterhouses. Josephine Pope of the WHPA later graphically described the scene and declared that many of the women present almost became converts to vegetarianism on the spot. These information-gathering visits led to letter-writing campaigns, public protests before city councils, and occasionally successful bids for municipal regulation of slaughterhouses. More important to veterinarians, the WHPA had decided to urge the city to hire a corps of municipal meat inspectors beyond the four employed at the time. The profession's leaders saw veterinarians as natural candidates for newly developing public health jobs like these, and were careful to stress their interest in the scientific maintenance of public health to concerned women such as the members of the WHPA.[26]

Perhaps the most famous woman to agitate for reform of meat production in the United States was Caroline Bartlett Crane, a citizen of Kalamazoo, Michigan. In a career spanning the first 15 years of the twentieth century, Crane wrote and helped to pass local, state, and federal legislation designed to regulate slaughterhouse sanitation and meat inspection. Crane's interest in the topic had originally arisen from a course on housekeeping

INTERIOR OF A SLAUGHTERHOUSE, PHILADELPHIA, 1909. The "filthy hands and aprons," meat "exposed to dirt, insects, and rodents," and lack of ventilation that Caroline Crane and other meat inspection activists decried were features of many small urban slaughterhouses like this one. *Courtesy of City Photographer, "3251–3253 Westminster Avenue, 1/16/09," folder 1798, photo 5319, City of Philadelphia, Department of Records, City Archives, Philadelphia, Pennsylvania.*

that she outlined for the Michigan State Federation of Women's Clubs. When she could not locate an expert to lecture on meat inspection, Crane decided to research the topic and discuss it herself. She began to visit meat markets in town and "the little slaughterhouses located in all directions about a mile beyond the city limits" in the spring of 1902. As she later remembered, the slaughterhouses impressed her as so "horrible and nauseating" that she was greatly surprised to find city meat sellers "openly hostile to the suggestion that any reform was needed." She particularly remembered a butcher "whose feelings had been outraged by the invasion of women into matters which were 'none of their business'"; nevertheless, the matter became public business with an exposé published in the evening newspaper. Crane and her club realized the enormity of the trouble they had stirred up when the mayor, health officials, and physicians asked to attend the next meeting of the Kalamazoo Women's Club.[27]

The women's club defended their actions and argued that local slaughterhouses and markets would lose out to the mass-market Chicago meat packers without the support and confidence of consumers. "If there is anything belonging to woman's sphere, surely it is the wholesomeness and cleanliness of the food she sets before her family," wrote Crane in an open letter to meat dealers. Women did most of the meat buying, and could threaten to boycott butchers and dealers. "If we have to get our meat from a distance we will do it," Crane continued, "but if the butchers will effect what we ask, the Chicago packing houses will find a fast-diminishing field in this part of Michigan." This argument surely struck home with local butchers struggling to survive against the underselling tactics of large Chicago packinghouses, although they were still reluctant to support regulation. The Kalamazoo city council advised Crane that municipal regulation of slaughterhouses beyond the city limits, similar to the milk laws already in place, could not be accomplished without an act of the state legislature. Upon discovering that other Michigan cities had problems similar to Kalamazoo's, Crane decided to pursue her cause at the state level. Finding stiff opposition, Crane and her colleagues waited another year for a change in administration and eventually got their bill passed with the help of sympathetic articles in newspapers and letter-writing campaigns. The women's experience with grassroots politicking reached other states and municipalities through the nationwide General Federation of Women's Clubs, "the women finding it easy, once one state had adopted it, to persuade another to do the same."[28]

Veterinarians played varied roles in Crane's experience. She had been befriended by George R. White, the city veterinarian of Nashville, Tennessee, while searching for model municipal meat-inspection statutes in 1902. A. S. Wheeler, the veterinarian at Biltmore, the Vanderbilt estate in North Carolina, also provided technical information and support. Crane had thoroughly studied federal meat inspection procedures with BAI Chicago inspector O. E. Dyson. Dyson had provided Crane with federal specifications for meat inspection and a copy of an official BAI address given to the American Public Health Association in 1901; she was greatly impressed.[29]

Upon returning home to campaign for the Michigan state bill, however, local veterinarians and their interests obstructed her path at every turn. A group of veterinarians had presented a rival bill which, according to Crane, "failed [and, she thought, was not even designed] to safeguard public health so much as to create numerous jobs for veterinarians." Crane claimed she would have "been so glad to join forces with them if it had been a good bill." Whether the trouble was political or practical, health reformers and veterinarians often did have difficulty combining their goals. While reformers accused veterinarians of concern with employment rather than public health, veterinarians found it "strange that the public is so slow in supporting and encouraging what veterinary science is capable of doing for the health of the people themselves." Municipal meat inspection should fall under the jurisdiction of local boards of health, argued veterinarian William H. Lowe in 1901, but "competent veterinarians must be selected to do this work if it is to be of any real value to the public."[30]

To become health officials, however, veterinarians had to join forces with people like Caroline Crane as participants in the political processes that created pure food legislation. Although they were often reluctant to do so, veterinarians eventually learned how to reconcile their scientific and professional goals with local, state, and federal regulation of food-animal production. Accomplishing this without alienating livestock producers required some careful maneuvering, as Pennsylvania veterinarians working toward state regulation of animal disease would discover.

In 1895, Pennsylvania veterinarians began work on an ambitious creation: the formation of a European-style sanitary regulation system for food products of animal origin. With Germany as the model, such a system included state surveillance and control (and funding) of methods of animal production and food processing. The Pennsylvania system needed state legislation to create a livestock sanitary commission and the office of state veterinarian, which would bear responsibility not only for regulation but also for investigation and research of animal diseases. The legislation required lobbying, and the Pennsylvania Veterinary Medical Association (PVMA) led the effort. The PVMA was composed largely of graduate veterinarians interested in promoting the new sanitary regulation system and their profession's role in it. Although the bills establishing the livestock sanitary commission and state veterinarian's office passed quickly, PVMA veterinarians had much more difficulty guiding the political appointments necessary to fill the newly created positions. Livestock producers' interests carried a great deal of weight with legislators, and they nominated candidates who would champion their economic cause. While not unsympathetic to this, Pennsylvania veterinarians sought a candidate who would also pursue research and be willing to regulate livestock production in the interests of human health. The PVMA's political committee quickly realized that they needed to present a high-profile member of the veterinary profession who could serve as a viable political appointee. Early in 1895, PVMA members voted to endorse Leonard Pearson for the state veterinarian's job. Although the position was created by legislative mandate in April 1895, it was only after a long delay, during which the veterinary societies of the state continued to work for appointing power, that Governor D. H. Hastings appointed Pearson on January 1, 1896.[31]

By all accounts an extraordinarily persuasive man, brilliant thinker, and strong leader, Leonard Pearson was to be a major American importer of German ideas on a state-sponsored sanitary regulation and surveillance system for animals and animal products. Pearson's training, which emphasized European bacteriology and pathology, was typical of that required to become a leader of the American veterinary profession at the turn of the century. In 1890, as a young graduate of the veterinary school at the University of Pennsylvania, Pearson spent a year in Germany. While there, he attended lectures in the veterinary schools of Dresden and Berlin, spent time working in the laboratory of the German army, and studied bacteriology in Robert Koch's laboratory. Pearson also scrutinized the organization of the German state's system of meat inspection, a branch of the "science of animal hygiene, correlated to the public health on the one hand, and to industrialism on the other." According to his friend of these student days, Alonzo Taylor, this system was developed to a state of high perfection at the time of Pearson's period of study there, and it fascinated the young student. Pearson imbibed other ideas as well: in 1890, he had watched Koch perfect tuberculin; in 1892, at the age of 24, he was the first on the American continent to use tuberculin to test a herd of cattle for tuberculosis. Pearson quickly won a reputation as a sharp and well-connected scientist. The Pennsylvania veterinary profession had found an effective leader, despite his youth. Pearson's success as a popular and respected veterinary scientist was reflected in his reappointment to the state veterinarian position by three succeeding governors before his death.[32]

Bovine tuberculosis, with its controversial bacteriological and practical implications, especially interested Pearson. His activities fit into a succession of international scientific investigations of this dread disease: Robert Koch had identified the organism in 1882 and worked with tuberculin in 1890. Koch, Pearson, and other influential scientists were to be involved through 1910 in a debate over the importance of the disease's transmissibility from cattle to humans.[33] Pearson had brought the tuberculin test to American cattle in 1892, encouraged the Bureau of Animal Industry to manufacture and supply tuberculin in 1894, and until his death in 1909 worked on a state regulatory framework designed to eradicate the disease. The key to Pearson's scientific success with tuberculosis, however, lay with his ability to balance it against the economic needs of the market. The *American Veterinary Review*'s editors asserted that Pearson managed to get along with the state livestock sanitary board, all nonveterinarians, as well as public health associations because his "personality, backed up by wise provisions in the law—framed under the guidance of veterinarians—has enabled the veterinary idea to predominate in enforcement." The "veterinary idea" argued that veterinarians were the proper applied scientists to take charge of livestock diseases, and public concern over tuberculosis in milk and meat animals occupied veterinarians more than any other single issue through the 1920s.[34]

The bovine tuberculosis eradication program, begun in 1896, was the keystone of Pennsylvania veterinarians' state-administered animal disease control system. Pearson received credit for designing and implementing the "Pennsylvania Plan" for eradicating bovine tuberculosis, and his design reflected his experience with the politics of regulation. He had also carefully studied the few existing eradication programs of other states and nations, incorporating their successful tactics. Pearson's rhetorical and practical strategies accomplished several goals simultaneously. Genuinely concerned with the public health implications of tuberculosis, Pearson seized the chance to use the issue as an entrée to more centralized state control over the health and mobility of domestic animals. This control turned on the utilization of veterinary expertise to administer the tuberculin test. Private-practice veterinarians in the field were signed on as district agents of the state system, thus linking local concerns with state regulation and providing additional income for qualified veterinarians. New institutions—a laboratory and experimental farm—eventually completed Pearson's ambitious plan. Finally, Pearson avoided alienating most livestock producers, an important base of his support, by making testing voluntary and securing legislative funds to reimburse farmers for condemned animals.[35]

The balance of agricultural and health concerns depended upon local conditions, Pearson argued. In areas with a high prevalence of bovine tuberculosis and poor inspection of meat and milk or none, eradication of bovine tuberculosis was a public health issue. Only in areas with adequate local inspection of the food supply did the eradication of bovine tuberculosis become primarily an agricultural (and therefore economic) issue. Pearson wrote that "since so much [had] been accomplished in the repression or eradication of other widely distributed disastrous diseases of animals," he had reason to believe that a veterinary-supervised tuberculosis eradication program would work. His antituberculosis Pennsylvania Plan was based in part on previous successful campaigns against bovine pleuropneumonia and glanders in horses.[36]

Explaining that the methods that depended upon the cooperation of the herd owner were the most successful, Pearson's system relied on the voluntary participation of livestock owners. The plan stipulated one compulsory component, that breeding animals coming into the state be tuberculin tested, but otherwise it relied upon voluntary requests from cattle owners to have their herds tested. If they were found positive, cows could be quarantined or condemned, with a reimbursement of $25 to $50 per animal (which was sometimes less than half an animal's value). Besides the indemnity, what could persuade owners to request testing? Pearson believed that producers wished to "escape the unceasing losses caused by the ravages of tuberculosis to their cattle," that they did not wish to spread tuberculosis among people ingesting infected milk and meat, and that their "neighbors frown upon the maintenance of a notoriously tubercular herd in their community." Finally, Pearson guessed that municipalities would increasingly regulate dairy products, and that owners of infected herds would not find a ready market for their products. Pearson's reasoning proved prescient. Under the new plan, cattle owners requested testing at a rate that exceeded the state's ability to oblige them.[37]

Pearson divided the state into districts and created a list of approved local veterinarians who could do the field work as their services were required. Access to tuberculin (the agent of the test), and the complex directions for its use, was available only to these field veterinarians through the state veterinarian, who also largely controlled the production of tuberculin. Hiring local practitioners reinforced acceptance of the tuberculin test by livestock owners. The district agent veterinarians also conducted public postmortems of reactive cattle, to visually demonstrate that the tuberculin test truly identified tubercular cattle.[38]

Pearson used his position as state veterinarian to exert even more expert control—while appearing to be permissive—over livestock raisers in Pennsylvania and other states. Cattle owners gained access to the desired state-funded testing and compensated slaughter program only through application to the state veterinarian. All tuberculin test results, as well as certificates on cattle imported into the state, had to be submitted to the state veterinarian, who approved disposal of the cattle. Moreover, since Pearson would accept import certificates only from approved agents of other states, his actions encouraged neighboring states to set up a district inspection system similar to that of the Pennsylvania Plan. The Pennsylvania Plan thereby helped veterinarians gain important roles as local mediators between the state and cattle owners, secured the tool of tuberculin testing solely for the use of the veterinary profession, and solidified state control over animal health.

The advantages for veterinarians were not lost on the profession around the nation. A 1904 *American Veterinary Review* editorial praised the state regulatory systems of Pennsylvania, Montana, and Minnesota. "Such instances should be an inspiration for the profession in other States," it admonished, "and they should strive to secure the great benefits that are bound to flow from such conditions." While the compensation in Pennsylvania was not much—a veterinary inspector made $5 a day, half what he could earn taking private calls—it was a steady source of income, especially for struggling young, scientifically trained practitioners and for those established practitioners losing horse patients. Although he did not forget his office's origins in the grassroots support of the Pennsylvania Veterinary Medical Association, Pearson nonetheless believed that his plan had something for everyone. Besides empowering veterinarians, it cleaned up the food supply and recompensed cattle owners fairly. According to Alonzo Taylor, Pearson felt that he had succeeded in "such reorganization and recasting of the German system as would adapt it to the social and political conditions of this country." His efforts were in fact usually aided by those social and political conditions; his Pennsylvania Plan became the basis for plans subsequently adopted in other states and for the general pattern of the U.S. animal health surveillance programs in effect today.[39]

At the federal level, veterinarians and their organizations also strove to develop their role as mediators between consumers and producers. At the beginning of the twentieth century, the BAI and its chief, Daniel E. Salmon, found itself in the middle of the debate over federal inspection of meat. Consumer "sentiment" demanded inspected meat, but if meat-packing companies paid for inspection, Salmon feared, consumers could expect to pay a disproportionately higher cost for meat. The answer to the problem lay with generous congressional appropriations for the BAI, which had managed to whittle the cost of examination by microscope down to 8.6 cents per carcass. Year after year, Salmon and the USDA argued for greater appropriations for the BAI on the grounds that its efficiency merited them, and that its workforce was too small to inspect all animals slaughtered in the United States for interstate and foreign commerce.[40]

Although the BAI's work had originally been concerned with economic losses on the international market for livestock producers and meat packers, its meat inspection mandate quickly became a domestic public health issue as well. After 1891, inspection of meat animals crossing interstate lines (before and after death) came under the BAI's purview, drastically increasing its domestic workload. In 1892, beef quarters inspected for consumption within the United States comprised 87 percent of total beef inspections, a shift in emphasis that made the BAI more accountable than any other federal agency for the healthfulness of the meat supply.[41] BAI inspectors routinely isolated and condemned animals with signs of diseases, such as Texas fever, that might be transmitted to other animals. Their most important function, however, lay with finding animals whose diseases could be transmitted to human consumers: septicemias (indicative of severe bacterial infection), tuberculosis, actinomycosis, and parasites, among others. Animals and animal carcasses showing signs of these diseases ended up as fertilizer rather than food, making much less profit for their owners.[42] The BAI's inspection work would soon become a major irritant to those in the meat animal business because the enforcement of regulations covering the mobility and disposition of domestic animals cut into profits. Located at the juncture of livestock interests, national politics, scientific research, and an intensified public demand for pure food, Salmon and the BAI were almost continually embroiled in debates over the procedures of federal meat inspection.

The 1891 Meat Inspection Act complicated the relationship between Salmon and the powerful businessmen controlling the domestic market for meat. By 1890, the "Big Five" meat packers, led by Philip Armour in Chicago, practically monopolized the slaughtering, packing, and distribution of mass-market meats. One historian has characterized the shape of meat production, from farm to table, as an hourglass, with the Chicago-based packing process making up the narrow waist. Since they were in control of the rate-limiting step, the packers could dictate the speed, distribution, and economics of the product.[43] Salmon and the BAI, on the other hand, felt that they had been given a mandate with few resources or punitive powers with which to enforce it. Packers had previously supported legislation that helped their export sales, but had resisted comprehensive domestic regulation. Regulation, after all, cost money; condemned carcasses were worth more as steaks than as fertilizer. The 1891 Meat Inspection Act was full of loopholes, even for the large operators who, as Salmon contended, benefited from the ability to advertise their meat as "U.S. Inspected."[44] When Salmon approached meat packers to make arrangements for federal oversight of disposal of condemned meat, he understood the weakness of his bargaining position. He resorted to under-the-table agreements with packers and slaughterers to ensure their cooperation rather than trying to enforce the ideal of regulating meat inspection.[45]

This action (and others) would later haunt Salmon, whose apparent political naiveté can be understood in light of his position and his priorities. Historians have described Salmon as a target for unfair criticism from all sides, including other veterinarians.[46] Head of an organization whose mandated purposes (livestock economics and public health) often conflicted, Salmon's continual involvement in controversy during his tenure at the BAI was almost assured. Moreover, his loyalty to his profession (veterinary medicine) added another dimension of urgency to his efforts on behalf of the BAI. The intimate connection between the BAI and the veterinary profession, Salmon implied, meant that negative press about the former amounted to a slur on the latter.[47] Salmon, who was perhaps the most visible veterinarian in the country, believed that his agency's impact on the success of the profession's goals was profound. Veterinarians therefore watched anxiously as the BAI weathered recurrent public scandals.

Problems with meat packing and meat inspection were becoming front-page news. Working together to regulate the ebbs and flows of the meat market, the Big Five meat-packing companies had more than nibbled at the edges of the Sherman Anti-Trust Act's prohibitions. By 1902, the Beef Trust (now six companies strong) was in flagrant violation of the act: members signed no-compete clauses, engaged in price fixing, and divided the country up into member-controlled wholesale districts. Beef packers earned the enmity of soldiers—future president Theodore Roosevelt among them—who had been forced to eat badly preserved, poor-quality meat during the Spanish-American War in 1898. The so-called embalmed beef scandal led to the punishment of the army commissary-general and, more important, to public scrutiny as the country formed its own opinion of federal corruption, corporate trusts, and the assurance of food quality.[48]

Journalists played an important role in keeping the public interest directed toward food quality as well as consumer fraud. Between 1900 and 1906, hundreds of articles about pure food in popular journals and newspapers stimulated consumer horror with such titles as "The Subtle Poisons"; even professional journal articles conjured fearful images with "Chicago: The Dark and Insanitary Premises Used for Slaughtering Cattle and Hogs." The most famous explosion of public concern over federal meat inspection began with three investigative articles in the British medical journal *Lancet*, written by English abattoir expert Adolphe Smith and published in 1905. In them, Smith outlined filthy conditions in the Chicago packinghouses, ridiculed meat inspections taking place at breakneck speed in badly lit rooms, and charged that the packinghouses employed workers infected with tuberculosis. Although the *Lancet*'s usual audience was limited to medical professionals, news of the articles spread quickly among interested laypeople in the United States. In the fall of 1905 Caroline Bartlett Crane heard about the *Lancet*'s attack on U.S. meat inspection. Crane's renewed interest in federal meat inspection dated to this revelation. Beginning in February 1905, Upton Sinclair published excerpts from his new novel (*The Jungle*) exposing packinghouse conditions in the *Appeal to Reason*, a socialist journal; at the same time, Charles Edward Russell's antitrust articles targeting the beef industry appeared in *Everybody's Maga-*

zine as excerpts from his forthcoming book, *The Greatest Trust in the World*. In March, a grand jury in Chicago brought indictments against the leading packing companies in the Beef Anti-Trust case, less than a year after the packers had successfully broken their workers' union by crushing a widespread strike. In April, the packers sought to calm fears about Chicago's meat by appealing to thousands of *Collier's* readers. Week by week, the controversy over the wholesomeness of American meat, and the fitness of large privately owned packing companies to supply it, escalated.[49]

The gathering storm over meat inspection was one of several incidents that troubled U.S. Secretary of Agriculture James A. Wilson over the summer of 1905. First, it had come to light that one of his employees in the plant industry division of the USDA had been successfully bribed by New York speculators to reveal inside knowledge about export cotton markets. The news shocked cotton dealers, and demands for Wilson's resignation reached Washington. "For eight years and five months nothing along the line of bad tendencies escaped me," he complained in August, "[and] finally the New York gamblers managed to corrupt one of our men." After asserting that "I shall stay with my job; my fighting blood is up," Wilson carefully probed every division of the USDA that summer. By September, he knew that Daniel Salmon had owned stock in a company secretly given an exclusive contract to supply the BAI with ink for stamping meat. Evidently this news, and the revelation that Salmon had made informal agreements with the Chicago packers about disposition of meat, had leaked outside the USDA and Salmon found himself at the center of controversy. "I personally thought that [Salmon's interest in the ink company] might be overlooked," Wilson wrote, "because I could not find any evidence of evil intention on his part." However, 55-year-old Salmon, who had directed the BAI since its founding in 1884, resigned amid public scrutiny on November 1, 1905. As Wilson put it, "I think he got tired of the whole matter and preferred to step out rather than be in the limelight of public criticism." Wilson asked Leonard Pearson, the Pennsylvania state veterinarian, to take the job; when he refused, BAI interim chief Alonzo Melvin assumed the position permanently.[50]

By mid-January 1906 the meat crisis had become overwhelming, and Wilson admitted that "these be ticklish times." A climate of secrecy and caution had taken hold at the BAI and the Department of Agriculture; public scrutiny of the meat supply had increased following the publication of excerpts from Sinclair's novel *The Jungle* in newspapers nationwide. This advance publicity by the book's publisher, Doubleday, concentrated on gruesome scenes of food production inside slaughterhouses as well as the horrors experienced by the working poor. Sinclair later said that he "aimed at the public's heart, and by accident I hit it in the stomach." His novel's socialist message was obscured as consumers repeatedly offended by tainted food, soiled politics, and graft exploded in moral indignation. In response, the USDA tried a desperate measure: on February 14, a "valentine" was sent to all packinghouses under inspection warning them to improve sanitary conditions—something that the USDA had no power to enforce, as everyone knew. By the end of February, Sinclair's novel had been published and was finding a wide audience. J. Ogden Armour, representing the packers, had a series of defensive articles ghostwritten and published in the *Saturday Evening Post* to little effect. The clamor reached President Roosevelt, who had his own reasons for detesting meat-packing companies, and the struggle was on to regulate the American meat industry in a manner that would satisfy concerned consumers (while not paralyzing the meat industry).[51]

Veterinarians felt unfairly singled out as scapegoats in the meat inspection crisis. Federal meat inspector and veterinarian D. Arthur Hughes kept a scrapbook that chronicled the turbulent year of 1906. In it he pasted anything he could find on the meat inspection controversy, and he wrote an uninhibited commentary that reveals the view of veterinary inspectors in the field. In one of his very first entries, Hughes responded to an article written by William K. Jaques, a Chicago physician and former city meat inspector. Jaques asserted that confusing federal inspection rules allowed parts of defective carcasses to pass; that inspection was ineffectual because it was hampered by graft, and that "scientific Germany" got the top-quality American meat, with the domestic trade getting second cut. Obviously stung by the insinuation that federal meat inspection rules lacked stringency and federal inspectors lacked scruples, Hughes wrote that "really the article is inspired by malice." Jaques' "egotism" made him "utter statements willfully which he must have known to be false," Hughes believed. Jaques' paper reverberated with a "spirit hateful to federal authority, with which presumably he had trouble when he held office in Chicago." Hughes concluded, "the truth is, Dr. Jaques had an ax to grind for he was in his paper flaying men whose power transcended his." Hughes' bitter comments reflected his vision of Jaques' paper and other similar essays as attacks on the morality of federal inspectors and the veterinary profession at large.[52]

Secretary Wilson also felt it necessary to defend the honor of his federal inspectors. "It has been reported that our inspectors are not vigilant enough," he wrote in May, but the real problem was that "I have found it impossible to get money enough to appoint as many inspectors as I think necessary to do the work completely." While exposés focused public blame on the BAI, Wilson also felt pressure from the president, whose marked personal attention to the meat controversy created bureaucratic discomfort at the USDA. "We occupy a disagreeable position in the Department," he complained. Through the tense months of May and June, the president and the packers postured around the shape of the incipient legislation. Three bills, covering meat inspection, railroads, and purity of food, matured together in Congress and each influenced the other's political path. With threats by Roosevelt balanced by minor changes in the bills, both the pure food act and the meat inspection law passed through Congress quickly at the end of June and were signed into law on June 30, 1906.[53]

The BAI, promised a greatly increased inspection appropriation, came out as well in that area as it could have hoped. However, the trouble with meat inspection—its high-profile political and sociocultural location—did not disappear for the BAI and the veterinary profession. As a political solution, the new meat inspection law remained limited. The BAI still lacked control over local slaughterhouses and had little power to punish violators of the new inspection act. These loopholes meant that the BAI still faced the risk of being made the scapegoat for inadequate meat inspection. Moreover, alarmed and defensive livestock interests ignored the chance to reform their own widely publicized procedures by blaming the bad publicity on journalists. "The public is being deluged with sensational articles; no charge seems to be too outrageous for irresponsible scribblers to bring forward," complained the *Breeder's Gazette* in mid-May.[54] All of these factors rendered meat inspection a somewhat dangerous proposition for BAI veterinarians. However, one of the underlying problems with meat inspection was perhaps far more profound. Veterinarians had misunderstood popular definitions of health, disease, purity, and unwholesomeness. The meat inspection crisis of 1906 taught veterinarians a new respect for the value of animal health among public health reformers and consumers of animal products.

Definitions of Health and Disease

Consumer concern had originated with the idea of ingesting milk and meat from only healthy animals. What constituted a "healthy" animal lay at the center of disagreements among consumers, producers, and inspection officials, even after passage of the 1906 meat inspection law. Inspectors and regulators had long understood that consumer "sentiment," not just science and sanitation, played a part in defining a healthy source of food. American consumers, unlike those in other countries, were loath to eat meat taken from a pregnant or parturient animal, or from animals less than a month old, for example. "The meat from such animals may not usually cause disease in the consumer," explained BAI chief Salmon in 1894, but the American people definitely felt a "repugnance in regard to it" that amounted to a moral prohibition on eating it. Therefore, consumers defined a "healthy" food-producing animal as one whose body would not be a source of physical disease or moral contamination. Such sentiments should be honored, Salmon agreed, "while there is an abundant supply of healthy animals" to be slaughtered for food.[55]

However, constant negotiation over the definition of "healthy" or "diseased" food products placed pressure on BAI veterinarians caught between the positions of livestock producers and consumers. Caroline Bartlett Crane, in a 1909 speech before the American Public Health Association, took up the matter of the "meat-inspection construction of the word 'diseased.'" Crane asserted that the BAI had redefined "diseased" in order to allow more meat to pass inspection (thus increasing producer profits). Particularly, Crane charged, parts of cattle and hogs infected with localized tuberculosis were being approved for human consumption. Crane quoted BAI chief Melvin, who had explained that "the word 'diseased' in connection with meat inspection has a meaning that differs from the generally accepted idea." Although eating meat from a diseased animal might be popularly thought abhorrent, some of the meat was in itself *not* diseased, and therefore not dangerous. Infected organs and meat should be condemned, but the rest of the carcass was fit for human consumption, argued Melvin. Crane disagreed, asserting that all meat from an animal with even mild, localized infection was unwholesome and should be discarded. She pointed out that veterinary inspectors could miss microscopic lesions of tuberculosis and other diseases, since they used only the naked eye to inspect carcasses. Were they really credible judges of wholesome meat under such conditions?[56]

Crane's suspicions reflected uncertainty within the BAI's corps of inspectors on how to judge the fitness of some carcasses for food. Internally, the BAI sought to decrease inspectors' confusion by sending out regular reminders of how certain carcasses should be labeled. Specific directives addressed the quick decisions inspectors had to make every day. Could the heart be passed if the thoracic lymph glands appeared diseased? How "slight" did intestinal lesions have to be in order for the viscera to pass inspection? In 1909, Chicago inspectors were assembled for a mandatory demonstration of the proper disposition of tubercular carcasses, "in order that greater uniformity would obtain in the various meat inspection centers upon this most important disease of food-producing animals."[57] Despite the criticism of consumer activists and the recognition of internal confusion, the BAI maintained that only "healthy" meat was passed by the federal inspectors. Regardless of the assurances of BAI veterinarians and other scientists, consumer advocates initiated several rounds of federal legislation over the next two decades that attempted to dictate the disposition of carcasses of infected animals.[58]

Milk, too, could be seen as "unclean" even if it carried no immediate bacteriological threat. Wholesome milk had to come from healthy cows. The presence of harmful bacteria in the cow or her milk, even if they were destroyed by pasteurization, rendered the milk undesirable. By 1921, health officials commented that "pasteurization of the bulk of the milk supply has outgrown the experimental stage and has come to stay." At the same time, the most extensive (and expensive) anti–bovine tuberculosis campaign ever conducted on milk cattle in the United States was under way, spearheaded by the BAI. The scientific community had accepted Theobald Smith's 1899 assertion that pasteurization killed all tuberculosis bacteria, thereby removing the fear of transmitting the disease to children through infected milk. Consumers, however, still felt that they and their families were ingesting "diseased" milk if it had come from tubercular cows.[59] As state and municipal regulations mandating tuberculin testing of cows became more common (around 1900), dairy producers began to cooperate with test-and-slaughter programs designed to eradicate bovine tuberculosis. Veterinarians used tools such as the tuberculin test to unite consumer concern and industry cooperation in order to define the criteria for animal health and product wholesomeness in local, state, and federal regulations.

In doing so, veterinarians assumed a broader public role and worked for acceptance of their profession as health experts and not just "horse doctors." Veterinary medicine was struggling to gain scientific and social recognition, to expand its employment base, and to influence national health policy without alienating livestock interests. The strategy helped the profession survive, but it also transformed it. Despite most veterinarians' orientation as independent practitioners, by the 1920s the majority of them had become state functionaries to some extent. Practitioners counted on employment as full or part-time food inspectors, vaccinators, and tuberculin testers with municipalities, states, the U.S. army, and the BAI. Because veterinary schools had restructured curricula around the laboratory sciences and food-animal husbandry, the intellectual orientation of the profession had shifted as well. Perhaps most important, veterinarians had tapped into the cultural value placed on animal health, and in doing so learned that they could shape their own role as expert arbitrators of the relations between Americans and their domestic animals.

CHAPTER FOUR

The Value in Numbers

Creating "Factory Farms"

at Midcentury

Consumers faced an ever-widening array of issues surrounding food in the mid-twentieth century. Concern over the healthfulness of milk and meat certainly did not disappear, but shortages, surpluses, and rationing surpassed all other worries during the years of the depression and World War II, and the products of animals' bodies remained much-desired premium foods. The tremendous expansion of large food-producing companies, processed food products, and food advertising influenced dietary guidelines, purchasing choices, and eating habits. The food market transformed itself into a largely national entity, with consumers at home, in military mess halls, and in restaurants eating ever-more-processed fare purchased from distant corporations bearing familiar brand names.[1] To enable these changes, food-production practices underwent equally profound transformations. The new circumstances under which animals lived satisfied values on which consumers, food processors, retailers, and animal producers could agree: food that met prevalent definitions of aesthetics, safety, and healthfulness that above all was cheap to produce and plentiful.

Animals and the processes used to turn them into food for human consumption became more and more invisible throughout the 1930s, World War II, and the prosperous years that followed.[2] Food-producing animals moved out of backyards and barnyards and into confinement houses, battery cages, and feedlots, the better to produce large amounts of meat, eggs, and milk at low cost. Concerns about the social and political roles of food outweighed the apprehensions consumers felt about animal health and welfare. Cheap and plentiful food could ensure survival through the years of the depression, provision the army, ameliorate the devastation of Europe following the war, and stave off Communism in hungry and discontented populations around the world.[3] Within this context, scientists working at experiment stations, land-grant colleges, and in commercial industries developed the technologies necessary to raise large numbers of animals cheaply and in less space. Veterinarians contributed to the trend toward intensive animal agriculture by working to reshape animals' bodily as well as external environments in the decades after the war. In the process, veterinary practitioners decreased their dependence on state employment and concentrated more on private populations of food-producing animals, whose value as individuals disappeared in the "production unit." Veterinary leaders redefined the rural practitioner's role as one of health advisor for herds of food-producing animals, rather than as physician for individuals. After the 1940s, veterinarians increasingly became part of the management team for the enlarging herds and flocks in the United States.

Beginning in the 1930s, transformations in the poultry industry heralded the development of health care for large numbers of animals. Not valuable enough as individuals to warrant veterinary attention, flocks of chickens and turkeys kept in ever-larger buildings could be annihilated by disease, threatening the producer's whole enterprise. Dealing with problems such as these helped the veterinary profession to expand in both numbers and scope after World War II. Finally, as the definition of disease took on economic as well as medical ramifications for producers, consumers began to question the moral issues inherent in valuing animals only in large numbers as sources of cheap food.

Disease Eradication before 1940

Of course, food-producing animals had been kept as herds rather than individuals prior to the 1940s. Indeed, approaching animal populations as large groups, referred to as "herd health," was a concept that had underlain the major animal disease eradication campaigns of the interwar years. These campaigns had usually (although not always) been characterized by the sacrifice of individual animals, no matter how valuable, that were infected with the target disease. By testing and slaughtering infected individuals, eradication officials believed that the health of the national population as a whole could be ensured. Two major eradication campaigns, for bovine tuberculosis and hog cholera, served as representative cases during the interwar years. The Bureau of Animal Industry oversaw both, providing materials, expertise, and direction of local efforts. The Federal-State Co-operative Plan for the Eradication of Bovine Tuberculosis began in 1917 with the blessing of consumers and continued through 1940, the year that the last two counties in the nation became accredited as tuberculosis free. The serum-and-virus vaccination method for hog cholera, tested in Iowa pigsties in 1908, initially overshadowed large-scale slaughtering as a way to eliminate that disease, but as the campaign continued through the 1950s, it refocused on the elimination of infected animals. Both of these campaigns took advantage of the fact that large numbers of practicing veterinarians were available to conduct them, on a local fee-for-work basis, around the nation.[4]

As their leaders had urged in the years approaching World War I, in the 1920s many practicing veterinarians around the country had turned their attention to eradicating disease in food-producing animals. Arthur Gold-haft, the Philadelphia veterinarian who moved to rural New Jersey, was not unique. A 1931 American Veterinary Medical Association survey of private practitioners, the majority of the profession, revealed that these veterinarians spent the largest segment of each day treating cattle, swine, poultry, and sheep, with little mention of horses. Survey authors carefully pointed out that the distribution of the American veterinary population in 1931 most closely matched that of cattle, not horses (as it had 30 years earlier), and that the amount of time expended on cattle and horses approximated their economic valuation. In 1935, veterinarians performed more than two million tuberculin tests on cattle per month nationwide. These statistics provided numerical validation for trends that veterinary leaders had been writing about in their professional journals: private practitioners had shifted their attention from horses to food-producing animals, and the hog cholera and bovine tuberculosis eradication campaigns had surely assisted this transition. During the most difficult years of the depression—when animals starved or were slaughtered and given away under relief pro-grams—rural veterinarians often found government-run disease eradica-tion programs to be the only available paid work.[5]

The eradication of bovine tuberculosis began on a national level with the formation of the BAI's tuberculosis eradication division in May 1917, which was funded by special congressional appropriations. Veterinarian

John Kiernan assumed the directorship of the division, and he immediately set out to sell his "area accreditation" plan of testing and slaughtering infected animals to veterinarians and livestock raisers, the two groups whose cooperation was most crucial. Not surprisingly, the former group proved the easier to enlist. Veterinarians had worked since the 1890s to establish themselves in states and localities as the official judges of dairy cows' tuberculosis status.[6] Under Kiernan's national eradication program, veterinarians were again uniquely authorized to administer the tuberculin test (supplied by the BAI), and thus to decide the fate of individual animals.[7] Kiernan attempted to enlist livestock producers by pointing out that they had the most to lose, in terms of money lost on salvage-slaughtered animals (especially valuable purebred cattle), quarantines and lower prices for cattle in an area with high rates of tuberculosis infection, and (most significantly) the loss of profits on bacilli-containing milk that could not be legally sold in city and town markets protected by inspection. As Leonard Pearson had done 20 years earlier in the Pennsylvania Plan, Kiernan sought to apply peer pressure among livestock raisers by excoriating the "conscienceless" among them who, "if unrestrained, would spread disease all over the country." His tactics of persuading livestock owners to be responsible for their herds' health reflected an initial shortage of supervisory officials—veterinarians—to enforce the eradication plan around the country. Kiernan wished to use the nationwide network of BAI inspectors, and BAI veterinary graduate and civil service examination records, to identify and hire local veterinary practitioners who would administer tests and condemn infected animals. He was soon able to do so with increasing congressional appropriations every year.[8]

Rural veterinary practitioners thus survived the interwar years largely as state functionaries for Kiernan's plan and others. As participants in these eradication campaigns, perhaps a greater percentage of veterinarians than members of any other profession worked for federal, state, and local governments in the 1930s. With about 10 percent fewer veterinarians in 1931 than in 1920, competition in the trenches of private practice had eased before the worst of the depression hit. Along with the fact that the profession had become "leaner" during the hard times of the 1920s, the employment that went along with the disease eradication campaigns helped rural veterinary practitioners survive the depression. Statistics are difficult to compile, but rural veterinarians who practiced during the 1920s and 1930s have consistently referred to government-funded disease eradication programs as their bread and butter. Many practitioners in the midwestern hog-producing states of Iowa and Illinois, despite competition from county extension agents, depended on serum-and-virus injections for hog cholera as their only steady work. Kansas veterinarian (later dean of the Kansas State Veterinary School) Ralph Dykstra described the bovine tuberculosis campaign as "engaging more of the personnel and more of the time of the veterinary profession than any other professional activity."[9] Local veterinarians served an important function for their government employers, also. Familiar with local zoogeography, they were unlikely to miss testing farmers with only a few animals—well beyond the 1920s, rural animals were still widely scattered in smaller groups.[10]

The tuberculosis eradication campaign increasingly became a victim of its own success by the end of the 1930s. Eleanor Roosevelt delighted veterinarians in January 1935 by releasing a syndicated article in which she praised the Bureau of Animal Industry for the success of the anti–bovine tuberculosis campaign. At the beginning of 1936, thirty-one states had gained entrance to the "tuberculosis-free honor roll"; by October 1940, the last two counties in the nation were declared accredited as tuberculosis free. Each accredited area meant the end of government-funded tuberculin testing work for the local veterinary practitioners. The more enterprising among them thought hard about how to return to the private sector completely. As they had already learned, only a dependable patient population, valuable enough to warrant professional veterinary attention, would ensure their individual professional success. As the political horizon darkened in Europe at the end of the 1930s, several factors (including their experience with disease eradication campaigns) led both rural veterinary practitioners and veterinary researchers to focus on larger-scale food-animal production.[11]

Despite the public health justification for eradicating tuberculosis, some veterinary professional leaders had already attempted to guide practitioners toward fulfilling the needs of agricultural producers rather than relying on public health campaigns. Louis A. Merillat, who edited the journal *Veterinary Medicine* in the 1920s and presided over the AVMA in 1925, boosted the cause of "animal husbandry" at the expense of public health work such as the tuberculosis eradication campaign. He warned practitioners that the sanitary campaigns were only temporary, and not to "risk a secure status in animal husbandry for a phantom in the domain of public health." After all, he reminded practitioners, "the producer is our friend who pays the bills."[12]

Of course, producers just as often tried to get out of paying the bills, or bypassed veterinarians altogether by utilizing home veterinary manuals, county extension agents, and speedily trained local "vaccinators" to treat sick animals and vaccinate against diseases such as hog cholera.[13] Nonetheless, Merillat (as had others before him) situated veterinary medicine solidly within the agricultural sector and disdained the comparative medicine and public health approach first championed in the late nineteenth century by Alexandre Liautard.

Merillat was doing more than simply dredging up the "agriculture versus public health" debate that lay at the heart of American veterinary medicine. He was responding to rapid changes in conditions. The impending end of large-scale disease eradication and vaccination programs meant that veterinary practitioners would have to rely less on government employment. The conclusion of the bovine tuberculosis campaign in 1940, and the end of the hog cholera vaccination program in 1953, created what veterinary historian O. H. V. Stalheim has characterized as "minor panics" in the ranks. The employment upon which rural veterinarians had depended for several years was disappearing, forcing them to reorient toward private commercial farm animal or pet-animal practice, or to leave the profession altogether. At the same time, livestock production and the security of the United States seemed inextricably linked at the end of the 1930s. "Food will win the war" had been the rallying cry of home-front efforts during World War I; and even in peacetime, the AVMA reminded veterinarians of their patriotic duty. The veterinary profession, J. S. Koen wrote in 1935, served as "one of the nation's greatest *defense* units . . . a civilian army of trained veterinarians in the Bureau of Animal Industry and among practitioners . . . protect[s] its livestock [and] furnish[es] the needed supply of meat." As the turmoil in Europe increased, Koen reminded his readers that every veterinarian represented a "cog in the national defense" by contributing to an adequate food supply and the profitability of livestock production. Rural veterinary practitioners, he implied, occupied a strategically important position as they manned the front lines of livestock production and meat inspection in the 1930s.[14]

Joining the Campaign to Feed the World

Fortunately for veterinary practitioners, in the 1930s tools were becoming available that promised to revolutionize their daily work. New chemotherapeutic agents appeared as treatments for common animal infections, and they gained early reputations as miracle workers. Sulfanilamide, commonly used in a preparation called prontonsil, attacked streptococcal bacteria—often the causative agent of mastitis in dairy cattle, a common and difficult veterinary problem. Veterinary bacteriologist William Hagan later characterized the early career of sulfas in veterinary medicine as a brilliant history as therapeutic agents. More was to come, however; a group of pathologists and other researchers, including Ernst Chain, reported in 1940 that the *Streptococci* and other bacteria were susceptible to the action of penicillin (first described by Alexander Fleming in 1929). Other antibiotics, including streptomycin, Aureomycin, Chloromycetin, and bacitracin, would prove their usefulness in the years following World War II. Veterinarians experimented with combinations of antibiotics, or mixtures of antibiotics and sulfonamides, to treat everything from wound infections to brucellosis. Sulfonamides continued to be widely used, since antibiotics could present problems of expense, bacterial resistance, complex dosage schedules, and toxicity. By the early 1950s, the availability of these and other chemotherapeutics had completely changed the character of daily veterinary practice, on the farm and in the hospital. These new tools would also raise veterinarians' expectations for their ability to treat sick animals. *Veterinary Medicine* reported to its readers that "it is freely predicted that within a few years chemical or antibiotic antagonists will be developed for all pathogenic organisms."[15]

In the mid-1940s, pharmaceutical companies, including Lederle Laboratories, Parke-Davis, and Merck, began manufacturing antimicrobial agents in forms usable by (and advertised to) veterinarians. For the busy veterinarian on the road, Parke-Davis's Penicillin-G did not require refrigeration. Abbott advertised its "crystalline Penicillin G" "for patients large and small . . . it doesn't matter whether the animal you are treating with penicillin weighs a pound or a ton." Commercial Solvents reminded veterinarians that penicillin had only recently been a laboratory curiosity, but was now necessary to veterinary practice; and the company mailed a free veterinary penicillin therapeutic reference table to every U.S. veterinarian. However, the cost of penicillin and other antibiotics, and suspicions about their spectacular reputations, prevented the more conservative rural veterinarians from embracing them immediately.[16]

Along with pharmaceutical companies, veterinary journals sought to reassure their readers about how these drugs could be productively and practically used for specific conditions and diseases. Veterinarians working in research laboratories, institutes, and divisions of a number of pharmaceutical companies collaborated with Selman Waksman and Herminie Kitchen on a lengthy *Journal of the American Veterinary Medical Association* article instructing veterinarians on how to use streptomycin and neomycin in their practices. Wound sepsis, mastitis, enteric infections—all once frustrating problems—could now be more readily controlled with penicillin and its counterpart, streptomycin. By all accounts, these efforts worked with veterinary practitioners; the use of antibiotics, especially as the price fell in the early 1950s, magnified in veterinary medicine. As veterinarian F. W. Schofield put it, "The modern medicine man often seems to need but two types of pills—antibiotics if the patient has a fever; vitamins with mineral if not." With their role in improving animals' health, these new chemotherapeutics fit well within a political and cultural framework that linked food production to national security in the late 1930s.[17]

Veterinarians did not escape the consequences, both positive and tragic, of World War II. The war had ratcheted up American livestock production, but had devastated that of Europe, with both trauma and disease taking their toll. William Hagan, then dean of the New York State Veterinary College at Cornell, was sent to Germany in 1945 to reorganize the veterinary service there and consult on animal disease problems. This was purely an administrative and not research position, and Hagan's hastily conferred rank of colonel in the U.S. army gave him the authority to conduct the usual veterinary measures against disease outbreaks—quarantines, testing, and slaughter. Returning to the United States, Hagan described to colleagues his observations of Germans freezing and starving to death in the bombed-out houses of Berlin.[18] Poor weather, chaos, and labor shortages contributed to desperate shortages of grain in Europe in 1945 and 1946; the Philippines had also suffered drastic depletions of livestock. With little food for people, animals received even less, and livestock production suffered.

U.S. feed production took on even more importance under these conditions, and livestock and feed industry journals warned American producers that feed would be short if domestic grain production faltered. All of these concerns worked to reinforce the USDA's usual urging for American farmers to become more efficient and produce more. *Poultry Tribune* quoted Secretary of Agriculture Clinton P. Anderson, who asserted that "we still have our own people to feed, including the military forces . . . at the same time, we are not forgetting our allies who now face hunger. [Agricultural] production played a big part in winning the war. . . . [I] suggest continued high production." Editorialists agreed that livestock producers had "a responsibility to the world's people who are short not only [on] feed, but also life-sustaining food"; under these conditions, waste was "intolerable." And food production had to address another threat—Communism. Livestock raisers were "the salesmen of democracy"; full bellies would prevent the people of western and central Europe from concluding that "democratic government does not have an adequate interest in their needs."[19]

Veterinary leaders intended to address these needs while expanding their profession in numbers and scope. Conditions during the immediate aftermath of the war encouraged more people to become veterinarians: livestock prices were initially high, other types of practice were expanding in the postwar economy, and the federal government provided educational benefits for returning servicemen. With a potentially large student body, and the availability of surplus war equipment and funds for building, at the end of the war seven new veterinary colleges sprang up in California, Georgia, Illinois, Minnesota, Missouri, and Oklahoma, and "one for Negro students" at Tuskegee Institute. These new schools, and their graduates, did not seem to present a threat to existing members of the profession, since numbers were still low because of the lean years of the 1920s and early 1930s. The shortage of veterinary personnel had become acutely apparent at the beginning of the war. Veterinary leaders believed that their small numbers led to a certain lack of influence in protecting their professional turf from pharmaceutical companies, county agents, and other competitors. Nonetheless, they recognized that the postwar era carried great responsibility and great promise for veterinary researchers and practitioners. Given educational opportunities and tools such as chemotherapeutics, vaccines, and nutritional analyses, veterinarians joined other scientists in addressing the challenges of feeding urban America and a war-torn world.[20]

Answering these challenges required the participation of an increasingly diversified group of scientists, all oriented toward efficiency in animal production. Animal husbandry became "animal science" in the interwar years, a specialty with its own university instructional programs, journal, and national organization. Species-oriented organizations, including the American Dairy Science Association and the American Poultry Science Association, formed in the 1920s and grew substantially in membership. These organizations also linked themselves to scientists at agricultural experiment stations and in universities. Through the 1920s and 1930s, the Bureau of Animal Industry continued to focus research on animal parasites and diseases such as poultry encephalomyelitis and brucellosis. The interwar growth of organizations and research in animal-oriented applied sciences, well described by historian Margaret Rossiter, also helped move animal production out of backyards and pastures and into isolated, large-scale, confined operations.[21]

The mid-twentieth-century disappearance of food-producing animals from backyards into large production units was only the latest of a series of similar agricultural reforms. Commercial orcharding; rice, cotton, and sugar cultivation; bonanza wheat farms; tree farms; and many other enterprises had instilled certain lessons in American agricultural economics and practice. First, larger-scale production units yielded more profit on a commodity with a tight profit margin; they also rendered animals who were worthless individually (such as poultry) valuable as a group. Second, monoculture, or the production of just one type of animal or crop, reduced overhead and labor costs. Third, even a partial, or regional, monopoly on products meant that consumer markets could be controlled to a higher degree. Fourth, vertical integration (ownership of all steps of production) was the most profit-oriented way to combine all of the above attributes under the purview of a single business operation.[22]

Animals were not strangers to living in large, monolithic populations, of course. In the horse's heyday, the largest urban stables held hundreds of animals; horses jostled each other on city streets while at their daily work. For cattle, sheep, and swine, the drives and holding pens of the nineteenth century had slowly given way to even more tightly packed railroad cars and stockyards. Certainly the development of packinghouse complexes in places such as Chicago at the end of the nineteenth century brought many innovations of efficiency and scale to the processing of animals. But the huge operations in which food-producing animals increasingly lived their whole lives in the decades after World War II had no exact precedent. In the process of meeting articulated goals for world food production (and sometimes unarticulated ones for increased profits), the development of these production units changed animals' living conditions and their very bodies and life cycles. In doing so, they helped to redefine Americans' valuation of food-producing animals and their products in the postwar era.

The New Poultry Industry

The poultry industry provided a salient example. The Republican Party's 1928 campaign promise of a chicken in every pot reflected the widespread understanding of chicken meat as a luxury food at the beginning of the depression. Eggs, at between 15 and 22 cents a dozen during the 1930s, represented a somewhat more affordable but still not abundant source of protein. By the end of the 1950s, however, both eggs and meat from poultry had become cheap staples of even a frugal American diet. Featured in cookbooks, articles, and the "Susan" food preparation series in *Good Housekeeping*, poultry and eggs were commonplace choices in recommendations for the postwar American table. Between 1935 and 1954, per capita chicken consumption rose from 18 to 30 pounds a year. Despite feed shortages and troublesome government regulations, poultry production continued to increase exponentially and prices continued an overall adjusted decline; an item that was a luxury in the late 1920s had become a dietary essential 20 years later.[23]

Cultural and social factors played important roles in effecting this transformation, of course, but they were contingent on efficient production of poultry products, and that began with changing the chicken's capabilities. Between 1935 and 1954, egg production per hen leaped from 122 to 183 eggs annually. It had required 12.3 pounds of feed and 89 days to produce a 3-pound depression-era broiler; by 1954, only 10.2 pounds of feed and 72 days were required. Changing the chicken was the prerequisite for cheap and plentiful poultry products, but these alterations were far from inevitable. The immediate postwar period represented a potential crisis for poultry production; it was a make-or-break time because of war-related overproduction, scarcity and cost of feed for birds, disease outbreaks, the need to educate consumers to purchase poultry products, and federally controlled supplies and prices. Poultry producers gambled as they followed expert recommendations to abandon traditional production methods and raise "improved" chickens. With the pharmaceutical industry for animals scaling up in tandem with the poultry industry, producers watched anxiously to determine the cost-effectiveness of the new antibiotics. Manipulating the bodies of chickens, and pushing the industry toward greater size and profitability, depended on cooperation among veterinarians, animal scientists, pharmaceutical companies, livestock producers, and feed companies. These new partners would help determine the future shape of the poultry industry, and the lessons learned there would also be applied to the large-scale production of other food-producing animals as well.[24]

Chickens proved to be biologically malleable, and thus good candidates for conversion into high-production, confined animals. Small, relatively easy to reproduce artificially, and valuable only in large groups, chickens for both meat and egg-laying purposes became targets of reform in the 1920s and 1930s (these were certainly not the first attempts to create a "better" chicken, but probably the most successful in raising each bird's food production capacity). The ideal chicken had to be genetically selected: the goal was a breed or type that laid eggs reliably or produced large muscles for meat and was not high strung and therefore amenable to confinement. The white Leghorn proved to satisfy animal breeders seeking high egg production and quiet temperament. Certain larger breeds, such as the Wyandotte and white Plymouth Rock, performed well as broilers (meat birds). In raising broilers especially, a slender profit margin made the weight-gaining abilities of the confined birds crucial to economic success. Thus, in the 1940s poultry producers consistently sought a larger, faster-growing type of bird.[25]

Poultry producers and scientists worked together to transform chickens into controlled "machines" for the production of cheap food. The artificial incubator had already restructured chickens' lives and the poultry business by the end of the 1930s. Most poultry producers purchased chicks from specialized hatcheries rather than hatching their own animals. Incubators, the "great hatching machines," had "almost made the mothering hen an anachronism"; hens instead spent their days producing eggs for sale. To produce maximum amounts of meat or eggs, of course, these birds had to be fed carefully. Traditional practices of feeding chickens table scraps and hand-broadcast grain had no place in scientific regimens that emphasized nutritional analysis for optimal egg and meat production. Nutritional science, a relatively new offshoot of agricultural chemistry, contributed research-based recommendations for bulk poultry rations. Vitamins, a relatively recent discovery, joined proteins, carbohydrates, and fats as important components of rations tailored to the ever-growing numbers of egg layers or broilers in each production unit. Feed concerned the producer as well because it was the single highest cost of raising broilers and producing eggs. Although feed could be produced most cheaply on the farm, the complex requirements of the "scientific" diet meant that in the 1940s producers relied more and more on commercial feeds to stimulate maximal production from their birds.[26]

Maximal production and profit also depended upon confining large numbers of birds and controlling their environment rigidly. Developments in housing practices enabled the tremendous expansion of total egg and meat production in the decades before and after World War II. For example, poultry raisers knew that hens would lay more eggs when they were exposed to more hours of light. Housing the hens indoors in completely controlled environments enabled producers to supply the extra light artificially. Thus they tricked hens into believing that summer lasted all year, stimulating extra egg production from each animal. Hens also lived in closer proximity to their sisters, to the extent that most poultry confinement houses were kept warm in winter entirely by the animals' body heat. When buildings were full of animals, producers saved money by avoiding the energy costs of providing heat. The whole management system's profitability, as *Poultry Tribune* advised in 1947, depended upon "a high average house capacity of laying hens the year around."[27]

However, diseases and behavioral problems threatened the large-scale poultry production unit in which chickens were treated as a flock. Some chickens reacted violently to the overcrowded conditions of confinement houses. Closely confined hens attacked and cannibalized one another, leading veterinarians and animal scientists to recommend a "debeaking" procedure that removed the sharp point of the upper beak. When faced with disease problems fostered by the animals' close confinement, poultry producers turned to veterinarians. Locally, veterinarian Arthur Goldhaft developed a lucrative practice and vaccine-production business serving New Jersey poultry producers. Federally, Bureau of Animal Industry officials agreed that confinement-house disease problems required taking a "herd" approach to poultry raising. Not surprisingly, they fell back on earlier experiences with livestock disease eradication programs. The BAI cooperated in the development of the USDA's National Poultry Improvement Plan, adopted in July 1935. The major objectives of this plan, as poultry raisers were informed by their journals, included the control of the highly infectious pullorum disease. This intestinal infection, passed from the hen to the chick through the egg, could wipe out large numbers of chicks within 10 days of hatching. The large hatcheries jumped at the chance to have veterinarians test their hens, and those raising younger animals with older ones in confinement houses were warned to do the same. With no medical remedy to cure it in a flock, pullorum disease fit into the old test-and-slaughter eradication model with which veterinarians were so familiar.[28]

However, other diseases that threatened poultry raised in large-scale confinement jeopardized veterinary control in a different way. Many infectious diseases proved to be susceptible to the new chemotherapeutic tools being developed to fight them, but pharmaceutical companies often angered veterinarians by selling these drugs directly to farmers. Sulfa drugs set the pattern of chemotherapeutic insurance against large-scale death in confinement poultry houses. By 1946, Lederle Laboratories regularly advertised its brand of sulfathiazole in poultry raisers' journals as a preventive for colds in laying hens. To the dismay of veterinarians, the formulations of sulfathiazole and related drugs made them easy to administer; the farmer could simply add them in specific amounts to the animals' feed. Pharmaceutical companies anxious to develop the market made sulfa drugs available to anyone, thus bypassing veterinarians' professional control.[29]

The new medications targeted specific diseases of the crowded poultry house. Coccidiosis, an enteric bacterial disease, had long been a major veterinary concern for broiler raisers. Traditionally dealt with by improving sanitation and culling infected birds, coccidiosis could wipe out most of a flock while leaving behind survivors that had immunity to future infections. By the end of World War II, Lederle's sulfaguanidine was readily available to poultry raisers who wanted to prevent coccidiosis outbreaks. In 1948, Merck Sharp and Dohme Research Laboratories announced the development of sulfaquinoxaline as a new and more potent coccidiostat for poultry. As with sulfathiazole and sulfaguanidine, sulfaquinoxaline was designed as a feed additive. Feed manufacturers and poultry raisers, although slow at first to adopt the new chemicals, soon realized their advantages in restructuring chickens' lives. Within 10 years of sulfaquinoxaline's introduction, Merck researchers asserted that the new chemicals had "permitted the poultry industry to expand enormously without the limitations and hazards of uncontrolled coccidiosis." While veterinary researchers at the BAI, experiment stations, and in private industry worked eagerly to help develop new chemotherapeutics, they unwittingly created difficulties for their brethren practicing in the field. Pharmaceutical companies willing to mass produce chemotherapeutics for animal agriculture saw no reason to restrict their market by making the drugs available only to licensed veterinarians.[30]

Practitioners interested in expanding their role in poultry production viewed the widening use of sulfa drugs with dismay. Traditionally not interested in chickens because individual animals were not worth the price of medical care, veterinary practitioners were in the process of reevaluating the services they could offer by making the intellectual change to thinking of the value of the whole flock. Poultry producers did not yet think of veterinarians as preventive managers, although they sometimes called on practitioners to diagnose and treat disease outbreaks. In the 1940s, preventive medicine for the flock usually excluded the veterinary practitioner. Farmers vaccinated their own animals according to the directions published in poultry journals. The same journals, when featuring articles about poultry diseases, often neglected to advise producers to call their veterinarian with questions or problems about diseases. The *Poultry Tribune's* monthly questions and answers section usually included at least one inquiry about disease sent in by a poultry producer and a detailed reply that contained medical advice. Clearly, in the 1940s poultry producers were not often getting their information, or their services, from their local veterinarians.[31] Veterinarians interested in getting into this market had to make two important changes: they had to see and advertise themselves as experts in flock or herd health (recognizing the value in numbers of animals); and they needed to learn how to use (if not control) the new tools, especially new pharmaceuticals, becoming widely available and of interest to poultry producers.[32]

Certainly opportunities existed for veterinary practitioners to be more involved with poultry production in the years immediately after the war. Arthur Goldhaft, whose poultry practice and vaccine research dated back to the late 1920s, traveled around the United States and the world promoting the responsibility of the veterinary profession to the poultry industry.[33] Even pharmaceutical companies, which had bypassed veterinarians by marketing sulfa drugs as feed additives, sought to make amends and ensure a partnership with veterinary practitioners. Sulfaquinoxaline, Merck's best-selling anticoccidial, provided a good example. Perhaps just as important as its action in completely preventing coccidiosis, the dose of sulfaquinoxaline could be titrated to match the level of environmental contamination, thus allowing birds to be exposed to just enough coccidiosis to induce immunity but not die of the infection. This titration was a tricky business, dependent upon the environmental and climatic conditions, the virulence of the local coccidial organisms, prevalent management and sanitation practices, and the general health of the flock. Assessing these factors and finding the best way to use anticoccidials was, as Merck researchers implied, the expected role of the veterinarian.[34] Coccidiosis, a local disease, could only be appropriately prevented by attention to local conditions.

Local veterinarians also received some assistance with coccidiosis and other poultry health issues from their colleagues engaged in research and poultry industry development. Leading veterinarians interested in poultry, some of whom were pathologists associated with state diagnostic laboratories, worked throughout the 1950s to educate the rank and file about their changing roles in poultry production. In 1954, for example, the annual meeting of the American Veterinary Medical Association included a symposium on the role of veterinarians in the control of poultry diseases, including presentations on developing a poultry-oriented practice, working with the feed industry, and utilizing the diagnostic laboratory service. An editorial in the the AVMA's journal that year urged veterinarians to be partners with the feed industry, which they asserted had "by self-discipline and by fostering nutritional science . . . become a giant and ethical industry." The partnership would, it implied, cut down on the amount of unregulated "incorporation of medicinal substances into feeds," including the use of sulfonamides to combat coccidiosis in poultry, without bypassing veterinarians.[35]

In the 1940s veterinary researchers eagerly addressed problems of preventive medicine in poultry and worked on changing the bodies of chickens. Chickens' viral and neoplastic diseases lay at the cutting edge of pathology at this time. The chief of the Bureau of Animal Industry promised poultry raisers in 1946 that he would seek more funds from Congress for research on poultry health, in large part to support the bureau's work on a vaccine against the viral Newcastle disease. In its highly contagious respiratory form, Newcastle could wipe out young birds and severely curtail egg production among older ones. The request seemed vindicated, as the BAI announced to the agricultural community in mid-1947 that its pathologists had developed a successful vaccine. *Everybody's Poultry Magazine* featured veterinarians O. L. Osteen and W. A. Anderson as they inoculated chicken embryos in the egg, ground up the infected embryos, filtered the virus-infected material, and added formalin to create the vaccine. Although Newcastle control through vaccination remained tricky for several more years, poultry producers could at least credit veterinary researchers for making progress on the problem. The other role of veterinarians lay with increasing inspection of poultry and poultry products. In 1936, only 12 veterinarians had been employed in USDA poultry inspection; by 1947, that number had increased to 153 and the volume of poultry inspected increased by 800 percent (even though inspection was not mandatory at processing plants).[36]

Rural veterinary practitioners initially expressed unease at the changes implied by researchers' new tools and producers' new management practices. The evolving emphasis on managing subclinical conditions, such as immunizing chickens against coccidiosis, placed daily veterinary practice on unfamiliar ground. Although rural veterinarians were not opposed to the idea of preventive medicine rather than therapeutics, they had been accustomed to procedures that addressed the health of ill or injured individuals valuable enough to justify professional medical attention. The narratives of veterinary practice, a crucial window into practitioners' view of themselves, revolved around the daily encounters between vets and individual animals. As veterinarian E. A. Schmoker recounted in a 1947 *Veterinary Medicine* story, "practice" meant the vagaries of lying on muddy ground in the winter, trying to replace a cow's prolapsed uterus.[37] Discarding their hypodermics and outdoor coveralls, changing to a suit jacket, and evaluating producers' daily management techniques represented a tremendous change in rural veterinarians' daily work and in how they viewed themselves. Nonetheless, their clients—livestock producers—increasingly had new tools available and new expectations for veterinarians within a social mandate to produce more food, more cheaply.

Cheap, Plentiful Food

The partnership between producers and scientists was perhaps best demonstrated by the 1950s' single most important development in animal production: the discovery and application of the antibiotic growth effect. Interest in "defense production" during the Korean conflict provided a further rationale for keeping efficiency high, but profitability was producers' main concern—despite the assertions of the *Poultry Tribune* that "the soldiers, sailors, and marines can't think about profit." Both goals were served by increasing each chicken's growth per day, whether as a young egg layer or as a meat bird, for faster maturation and cheaper feed costs per animal. For broiler producers especially, each added week of growth meant up to 10 cents of extra feed cost per bird, an increase that could wipe out the margin of profit and mean a net loss to the producer. Vitamins, hormones, and other additives had already established themselves as growth promoters and raised producers' expectations. By the early 1950s, the poultry industry was primed for reports of new compounds that would stimulate chickens' bodies to develop more rapidly.[38] Industry researchers worked steadily to identify these miracle drugs, but the next big breakthrough came from an entirely different, and surprising, source.

In 1948, nutritional biochemist Thomas Jukes and his colleagues were searching for microorganisms that would produce animal growth factors when they noticed that *Streptomyces auriofaciens* (used to make the antibiotic aureomycin) caused chicks to grow very quickly and thrive when it was fed to them. The researchers achieved even more spectacular results when they fed the by-products of aureomycin production to piglets. Jukes had been educated at the Ontario Agricultural College in Guelph, and he realized that the growth-stimulating effects could represent another important practical benefit of antibiotic compounds. These chemicals could be utilized not only as therapeutic agents in states of disease but also as enhancements to health in animals, and perhaps even in human beings. American children, still facing polio, poverty, and other unhealthful conditions, could benefit by having their "environment cleaned up" and their "disease level lowered" through antibiotic-enhanced food. Young food-producing animals were the obvious first target for testing, as veterinary and agricultural leaders immediately began linking the antibiotic growth effect to increased food production for the nation and the world.[39] So well publicized was this discovery that it appeared in popular magazines and political cartoons. Moreover, agricultural producers' interest in utilizing the antibiotic growth effect fit well into corporate reality: by-products from antibiotic production, previously a disposal problem, were incorporated into rations for growing chicks. Thus, the cost of the chemicals could be kept well below that of the perceived benefits of faster growth. By the end of the decade, residues containing antibiotics had become standard ingredients in feed for chickens, hogs, and cattle.[40]

Back in the laboratory, several research groups (including that of Thomas Jukes, who worked for Lederle Laboratories in the 1950s) searched for the physiological reasons that could explain the antibiotic growth effect. While the exact mechanism of growth stimulation eluded them, researchers did demonstrate that the antibiotics selectively killed intestinal flora. Animals fed subtherapeutic doses of antibiotics had different bacterial profiles and a reduction of toxic products in their intestines compared with untreated animals. Indeed, the antibiotic growth effect did not seem to work in "germ-free" animals living in sterile environments; conceivably, those animals already lacked harmful, growth-reducing bacteria. Researchers concluded that however the effect worked at the cellular level, in practice it induced changes in the intestinal flora of the chickens, pigs, and other animals on whom it was tested.[41] Anyone administering these growth promoters was, therefore, transforming the internal environment of the animals being fed the compounds.

Veterinarians lauded the subtherapeutic use of antibiotics as part of an appropriate preventive regimen for food-producing animals raised in confinement. The trend toward larger production units only amplified disease problems; indeed, disease posed the greatest threat to large-scale producers. As with sulfa drugs and coccidiosis, the constant use of antibiotics conditioned animals to survive in close confinement by altering their internal environments as a way to stave off disease. The use of antibiotics in feeds altered animals' external environments as well. Manure contained fewer amounts of some bacteria and more of others, leading to different concentrations in the cages, air, and on the floor of production facilities. Feed and water compositions changed also to produce stronger eggshells, larger breast muscles, faster weight gain, and better survivability in chickens. Chickens' bodies and environments had been transmuted into forms that would grow a large number of animals cheaply, the better to serve producers' profit needs and the value placed by consumers on inexpensive, wholesome food. And the strategy worked: in the early 1950s, Americans consumed up to eight times more poultry products than they had in the late 1920s. As long as eggs and poultry meat remained "a bargain," thus satisfying the consumer's first criterion, the sky seemed the limit. A *Poultry Tribune* commentator asked, "What is the limit on egg consumption?" and then answered his own question, "Who knows?"[42]

During the 1960s, animal scientists and veterinarians transplanted the model pioneered by poultry, including physical changes in animals' bodies and their environments, to pigs. Fed antibiotics and hormones, young pigs experienced much faster weight gain and a delayed onset of puberty. Adult females, now artificially inseminated, produced more litters per year with less rest between pregnancies; the number of pigs farrowed per litter increased from 6.8 in 1959 to 7.2 in 1969. Veterinarian George A. Young even pioneered a method of delivering all piglets by cesarean section, the better to produce germ-free animals (the ultimate preventive medicine). All of these bodily improvements meant that animals could be made into more efficient producers. Such animals raised in large numbers in confinement (a type of housing new to pigs) could bring higher levels of profitability, and swine producers abandoned traditional methods in droves during the 1960s. About 8,700 new hog confinement buildings were constructed in 1963 compared with almost 40,000 in 1973, mostly in the midwestern hog- and corn-belt states. Like chickens, pigs had become valuable only in large numbers.[43]

The transformation in animal agriculture in the 1960s meant further changes for many veterinary practitioners who had cared for food-producing animals. The practices of the previous generation, and even the training available in most veterinary colleges, no longer applied when clients wanted services for herds. Moreover, changes in veterinary professional leadership in the 1950s only contributed to practitioners' unease. The structure of the veterinary profession, in place for 70 years, received a blow with the dissolution of the Bureau of Animal Industry in 1953. The bureau had at various times provided crucial employment for veterinarians; it presided over the largest concentration of veterinary research, the national disease eradication programs, and meat inspection; and it supervised the development of veterinary education (see Chapter Two). The profession had depended upon the bureau for leadership and for influence in Washington. After its dissolution, bureau employees were reassigned to new agencies, and these regulatory and research veterinarians became more isolated. With the fragmentation of the bureau, veterinary medicine lost its concentration of federal influence and its consolidated leadership base in the midst of the animal industries' transition to large-scale confinement production. Once again, veterinary associations and colleges scrambled to create direction for their constituents within this upheaval, and they utilized veterinary journals to get the message to practitioners nationwide.[44]

Modern Veterinary Practice, the veterinary journal read most often by food-animal practitioners, compared the transition of the 1960s to the end of the disease eradication campaigns in the previous decades; each "contained the seeds of its own extinction."[45] According to veterinary leaders, the future of the agricultural practitioner depended on his ability to adapt to the new conditions. *Modern Veterinary Practice* warned its readers in 1965 that the coming years would hold fewer farms and more food-animal "factories," and fewer large-animal practitioners. Those practitioners would treat animals, valuable only in large numbers, from afar; diagnosis and preventive medicine would be the veterinarian's only job.

Confinement will reach new heights. The veterinarian working with such a food factory will function more as an engineer than a medical man, and will be more familiar with computers than stethoscopes. His work will concentrate on herd-wide disease prevention and growth planning rather than specific diseases, and he may go weeks or months at a time without ever seeing a live pig or cow. . . . The profession is going to change, change fast, and change extensively. The DVM who wants to keep practicing is going to have to change with it. He is going to have to avoid basing his practice on routine procedures that truly do not require his skills.[46]

"Routine procedures" had, of course, long been a bane to veterinary leaders anxious to claim a proprietary knowledge and skills base for their profession. As long as livestock owners could administer their own antibiotics and give their own vaccinations, veterinarians had experienced serious competition for some of their services. The updated motif of "herd health," with its emphasis on disease prevention and tight profit margins, promised salvation to struggling food-animal practitioners—if they could prove themselves useful to "the educated farmer."[47]

In order to expand their new role, veterinarians interested in herd health redefined disease to include a much wider array of animal management problems. Clinical disease problems in individual animals had always been the province of veterinarians. In addition, veterinary leaders argued that subclinical disease in its broadest sense should also be considered within the veterinary purview. Subclinical disease included such things as infertility in breeding animals, mastitis in dairy cows, and even growth curves in groups of young animals that did not meet the producer's expectations. Finally, no veterinarian interested in the health of the herd or flock would ignore inefficiencies in management. Such problems could lead directly to outbreaks of infections or loss of profits through low weight gain or other production parameters. Thus, "disease" came to encompass anything that food-animal practitioners could identify as preventing the herd from attaining its maximal production capacity—an economic rather than a medical definition, and one that applied only to animals valuable in large groups.[48]

Along with its explicit economic framework, the movement toward herd health also signified an important underlying change in attitudes toward food-producing animals. Consumers concerned with the healthfulness of animal-derived foods had for several decades associated animal health with animal welfare. A healthy body implied that the animal was fed and housed well and treated kindly. Thus healthful food was also humane food. After the implementation of the Meat Inspection Act of 1906 and numerous local milk inspection laws, and with the success of veterinarians' early disease eradication programs, consumers could relax their vigilance against diseased meat and milk. The purple USDA Approved stamp meant more, however, than a badge of wholesomeness; it also symbolized the clean living of the animals from whose bodies the ham or bottle of milk had come. With this assurance, consumers did not need to be anxious about the messy details of raising, slaughtering, and processing food-producing animals. All these processes could now be imagined as regulated, humane, and clean. Reality may not have matched imagination; as historian J. F. Smithcors has put it, "it is understandable that nice people would be offended by what necessarily goes on in even the best-run abattoir." Moreover, as Joanna Swabe has described, the "bucolic images" of food industry advertising replaced most peoples' daily experience with livestock production in the second half of the twentieth century. Raising, slaughtering, and processing animals had become invisible, with little or no consumer apprehension of animals' health or welfare.[49]

Stock raisers had also reevaluated their attitudes toward their animals. While veterinarians' clients had certainly always been interested in profit, they had also cared enough about the welfare of individual animals to secure professional medical care for them when necessary. Viewing animals as having value only in numbers, however, narrowed the scope of purposes for veterinarians' services. The need for herd health, *Modern Veterinary Practice* noted in 1969, had followed from changes in attitude on the part of the livestock owners. Owners now valued animals strictly as profit makers: "Once the farmer accepted the individual cow as an expendable commodity and not 'part of the family' he would take a second look before calling the DVM. These changes have taken place, and there's not much use in saying 'if something had been done ten years ago. . . .'" No doubt practitioners did look upon the past with some nostalgia in the face of upheaval. Beyond that, however, the journal's statements exposed two assumptions by practitioners: that the identity of their patients had changed and that some action on their part could somehow have prevented it. Food-animal practitioners met the "progress" of confinement-based animal production systems with ambivalence.[50]

Certainly the change in identity for food-producing animals that took place between the 1930s and the 1970s had far-reaching consequences for the animals themselves. If it was not cost-effective to treat an individual animal's affliction, then it did not get treated, no matter how severe it was. Along with individual value, animals also had lost autonomy and control over their environments, both internal and external. The host-parasite relationship between animals and intestinal bacteria and parasites (such as coccidia) had been completely redefined by producers, pharmaceutical companies, and veterinarians. The crowded conditions of controlled indoor environments removed individual animals' ability to respond to their surroundings.[51] Producers ignored normal behaviors or categorized them as deviances that humans should control. As members of production units, animals had lost much of their individual integrity. When overcrowded hens acted on their stress by cannibalizing one another, their keepers interpreted the behavior (not the environment) as the problem and removed the sharp ends of hens' beaks. If the ventilating fans—or belts conveying feed, or pumps circulating water—failed, then animals suffocated, went hungry, or thirsted in their cages and rooms.

If veterinarians pondered these ramifications of the transformation of food animals into expendable commodities, neither they nor their journals discussed it extensively. Nor were most American consumers concerned enough about these issues to stop purchasing meat, poultry, and eggs (although interest in animal welfare rose in the last three decades of the twentieth century), mostly because animals had moved into specialized facilities, out of sight. Animals' value in numbers supported the web of cultural beliefs, consumer desires, and corporate profit motives that made up the logic of the production unit. Veterinarians, while lamenting the transformation in their work practices and narratives dictated by this logic, nonetheless assisted it by participating in scientific research and accepting employment as herd health experts. The transformation in animal agriculture caused tremendous difficulties for veterinarians, however. At the least it required retooling intellectually and practically: feed analyses replaced physical examinations, and computers replaced syringes and hoof trimmers. To be sure, this change occurred slowly in most of the United States; most farmers did not have the capital to attain national stature as producers or even to develop a midlevel operation. But throughout the 1950s, '60s, and '70s, herds grew fewer and larger, and small farmers became livestock producers or ceased raising animals for profit. As *Modern Veterinary Practice* asserted in 1969, food-animal veterinarians could choose to become large-herd specialists or find another line of work: "In many communities the survival of veterinary services depends on redefining the veterinarian's role. And it can only be successfully redefined in terms of the farmer's wants and needs."[52]

Faced with these difficulties, some veterinarians converted to herd health after the mid-1960s; some scraped by with traditional practice; a few left veterinary medicine altogether; and many searched for a new patient population on which to concentrate. Like their forebears threatened by the decrease in horses' value a generation or two earlier, these practitioners sought a patient population whose owners valued individual animals enough to purchase medical care for them. With suburbanization rapidly encroaching on rural areas, the obvious choice was to concentrate on the newcomers' animals: pets. While most rural veterinarians had not entered their profession to minister to dogs, cats, and other companion animals, they were not without some experience in this area. Companion animals had been steadily increasing in numbers since the 1920s, both in Americans' homes and in veterinarians' offices. By 1965, with workhorses a dim memory and food-animal practice transformed, many veterinary practitioners turned to companion animals as the patient population that best guaranteed a bright future.[53]

CHAPTER FIVE

Pricing the Priceless Pet

Although Americans had always kept animals as companions, the status of pet keeping in American society and culture was transformed during the twentieth century: affection for companion animals became fully integrated into mass consumer culture.[1] Once an isolated luxury hobby, spending money on pets has become a widespread consumer activity. In the late 1990s, an estimated 129 million pets lived in 59 percent of all U.S. households, and Americans spent more than $12 billion ($5 billion on food, and $7 billion on veterinary care) every year on dogs alone.[2] Yet a hundred years ago, few Americans ever purchased the equivalent of today's pet foods, toys, or grooming services; even veterinarians spent almost none of their time treating pets. During the twentieth century, U.S. pet owners have become habituated to expressing their regard for their companion animals by buying pet-related products and services. Owners have translated their affection for their pets into economic value for animals that have no worth as sources of transport, food, or other products.

Veterinary researchers and practitioners have both contributed to this transformation and benefitted from it. Researchers created vaccinations for companion animals that veterinary practitioners integrated into annual examinations. By cooperating with other animal industries, such as pharmaceutical and pet food companies, veterinarians' office visits became an important component of proper care of the middle-class pet. Especially after World War II, veterinarians built hospitals for animals, incorporating technologies formerly available only in human hospitals. Creating higher standards of care for pets not only addressed animal and human health issues, but also capitalized on Americans' willingness to spend disposable income on animals that were increasingly viewed as part of the family. In the middle decades of the twentieth century, companion animals became practitioners' most stable patient population because veterinarians had helped to shape the market for pet care. However, this could occur only after veterinary practitioners learned to accept a new paradigm, valuing animals for sentimental rather than economic reasons. Veterinary leaders, too, redefined their profession in terms beyond their traditional characterization as scientists devoted to protecting the livestock economy and human health. From the 1930s on, veterinary leaders characterized their profession and its institutions as a "great humane society." Veterinarians put themselves forward as scientists who were also champions of animal welfare. This strategy served veterinarians well because it defined a higher moral purpose for their work and supported the valuation of animals based on sentiment.

Twentieth-century Americans incorporated the companionship and affection implicit in pet keeping into major social structures: mass popular and consumer culture. Seemingly unquantifiable, affection for animals nonetheless took on economic value through a particular set of historical circumstances. The Victorians, from the queen's example to the activities of animal protection groups, inaugurated widespread acceptance of sentiment toward pets in Anglo-American culture. After the Civil War, interest in animal protection, focusing on dogs and horses, took hold in the United States in the form of antivivisection and humane associations. Along with these organizations came, as historian Keith Thomas has described, the "modern sensibility" toward animals. This does not imply that Americans had not cared for companion animals before this time period; rather, the heavy imprint of Victorian culture cleared the way for the popular consumer interest that would manifest itself in the United States after World War I.[3]

Between World War I and the 1930s, companion animals became cultural figures of note as well as the subjects of growing commercial efforts. Dogs and other animals assumed roles as nationally popular characters in articles, books, and films. Narratives about these animals were certainly not an early twentieth-century invention, but they did not become widely popular until after World War I. Albert Payson Terhune, well-known author of twentieth-century American dog stories, remembered that he and his family had almost starved because he could not sell his animal stories to editors before the war. In 1914 he sold one of his first stories for $200; by 1923 an article of the same length brought $2,000. His first book of dog stories, published in 1919 only after a long search for a willing publisher, attained immediate popularity and went through more than forty editions in Terhune's lifetime.[4] The timing of Terhune's success illustrated the importance of World War I in bringing companion animals, especially dogs, into mainstream popular culture.

Dog stories owed their popularity after the war to widespread interest in canine military heroes. Articles about heroic dogs in World War I served their writers' purposes as dramatic vignettes designed to capitalize on Americans' romanticized views of animals and interest in the events of the war. War-dog stories assured families back home that dogs served not only as lifesavers for the soldiers in the trenches, but also as guardians of morality in an immoral situation. War threatened both the physical body and the mental and spiritual health of the soldier. Dogs could be heroic by virtue of their companionship alone. "Hatred against the brutal Huns did not destroy the otherwise affectionate disposition of our boys at the front" because of the many pets "the boys" adopted and lavished with affection. The affection and loyalty of the dogs meant that the soldiers' sense of humanitarianism would not fall victim to the horrors of war. The image of the companionate animal as a sentinel of morality, although not new, gained credence for adults as well as children in the war narratives.[5]

Narratives about dog heroes described civilian as well as military exploits; many popular post–World War I dog stories featured canines who had saved the lives of ordinary humans. Perhaps the most famous story of civilian canine heroism in the 1920s was that of Balto, Togo, and the dog teams that raced to save the town of Nome, Alaska. In January 1925, the isolated town was placed under quarantine because of the threat of a diphtheria epidemic. With cases of illness and death increasing, the town's physician desperately sought a shipment of antitoxin that was delayed at the railroad terminus in Nenana, 674 miles away. Poor weather prevented airplane travel, so the antitoxin had to be delivered by dogsled. Nome's plight attracted widespread media attention; newspapers documented events daily. "The world held its breath" while teams of dogs and drivers relayed the antitoxin through wind, extreme cold, and blizzards. Five days after beginning the journey, the last team pulled into Nome, delivered the antitoxin, and gained almost instant fame.[6] Balto, the lead dog of the finishing team, became the representative hero for Togo, Jack, Dixie, and the other lead dogs portrayed as fighting their way by instinct through the snow that blinded their drivers. The "endurance, fidelity, and intelligence" of the

Alaskan dog heroes was celebrated with a bronze statue of Balto erected in New York's Central Park.[7]

Heroic narratives like the Nome story helped to increase the popularity of dog stories and became a staple of motion pictures starring dogs. Movies featuring dog heroes, or "flea features" as they were fondly known in the trade, gained enormous popularity and fame for their stars beginning in the early 1920s. Rin-Tin-Tin exemplified the canine star: male, a German shepherd, a survivor of the war, enormously popular, and rich—Rin-Tin-Tin made $500 a week and his canine colleague, Peter the Great, almost $1,000. Rin-Tin-Tin, whose owner had formerly shared his mess kit and army blanket with the dog, received 10,000 fan letters every week and was insured for $100,000. Strongheart's fans "packed their belongings and their pets to make the grand tour" of theaters across the nation with him in 1924. In their movies, these dogs were shown attacking villains, exhibiting extreme loyalty to their masters, and heroically defending and assisting helpless people. Americans, especially children, were enthralled by the canine heroes, and dog fanciers everywhere basked in the glow of the animals' new importance.[8]

Stories and films thus played a crucial role in translating the working function of dogs into affective terms in postwar American popular culture. The dogs immortalized in epic narratives and films functioned as representatives of human cultural values, but they also came to exemplify the laudatory characteristics of all domesticated dogs. Dogs, "man's best friend," had shown themselves to be worthy companions for children and adults alike. Canine film heroes blended the ordinary with the fantasy, making their movies immediate and accessible to anyone who had a relationship with a dog. The family dog was, after all, cut from the same cloth as Rin-Tin-Tin, and carried the same noble impulses. Rin-Tin-Tin's successor as ruler of the canine movie stars was Lassie, a collie belonging to an ordinary little boy. What Lassie lacked in military credentials, she or he (there were several over the years) more than made up for with civilian gallantry and fierce loyalty to the young master. The dog next door was a hero too.

These positive popular culture images gave the family pet a new identity as a focus of consumer culture in the interwar decades. Consumerized pet keeping expanded because it tapped into popular culture and was a good focus for creating market opportunities. Pet animals themselves had little economic value, but their cultural representations, created in postwar stories and films, did. The human-animal bond immortalized in heroic narratives made pets fit subjects for their owners' solicitude. Appropriately channeled, owners' eagerness to care properly for their pets would stimulate the growth of pet-related industries. Pet food manufacturing, which expanded from very humble origins into a robust industry in the interwar period, provided a salient example. Prior to the war, pet and kennel owners formulated their own dog rations at home, using meat, dog biscuits if they could get them, and various additives. After the war, companies such as Spratt's and Ken-L Ration seized the opportunity to greatly expand the commercial markets for pet foods. Initially targeting kennel owners, companies quickly discovered the willingness of private pet owners to purchase prepared pet foods. In 1924, only about 10 companies produced pet food; 10 years later, the number had risen to 175 (including General Foods, Armour, and other companies that produced food for humans).[9]

Even the depression did not stifle the growth of consumer demand for pet products. While other nationwide industries collapsed, companies that produced pet foods claimed an expansion during the 1930s. The National Dog Food Manufacturers Association in 1934 asserted that sales had about doubled each year since the stock market crash. In 1934, $40 million was spent on dog food; by 1936, consumers spent more than $100 million on pet food, and the industry was the country's second-largest user of tin cans. Some portion of these purchases—estimated at up to one million pounds a year, or up to 20 percent of total production—was accounted for by human consumption of dog food during the 1930s. Congressional concern over this fact in 1936 led to federal meat inspection regulations being applied to dog foods (an action that the veterinary profession officially supported). Even allotting only 80 percent of dog food consumption to dogs, however, the numbers of pet animals having food purchased for them increased greatly during the depression. Moreover, the tremendous growth of dog food sales continued unabated even after consumers could afford to buy chicken or beef rather than pet food for the family table. Purchasing food for pets was becoming a widespread consumer practice, supporting a lucrative industry through the interwar period.[10]

The explosive growth of pet food sales depended in part on the industry's active efforts to link its products, through advertising, to the popular image of heroic dogs.[11] Pet food advertising used testimonials from breeders of champion dogs and veterinarians, and several companies entered into contracts with the owners of dog stars like Rin-Tin-Tin and Lassie to gain a competitive edge. One of the early prepared dog diets, Ken-L Ration, was marketed using the image of Rin-Tin-Tin.[12] Lassie's appearances in dog food advertisements spanned decades.[13] Magazine advertisements featuring famous dogs also teamed up with popular company-sponsored radio features. The dog stories that accompanied pet food advertisements captivated readers. Chappel Brothers, the company that produced Ken-L Ration and related products in the 1930s, sponsored "Rin-Tin-Tin Thrillers" every Thursday evening over the NBC radio network.[14] The Iowa-based company that produced Red Heart also sponsored the weekly Lassie show on NBC radio: "Hear MGM's famous collie star in her own thrilling, real-life stories every Saturday."[15] The popularity of dogs sold even non-pet-related products: a 1937 Schenley Whiskey Company ad featured the Wilken family dog, Jessie, and her new puppies. Forty-five thousand people wrote in to request a photograph of the dogs.[16]

Along with the reliance on famous dogs, pet food advertisements legitimated the affection owners felt for their pets and reminded them of their duties to care appropriately for them. Proper care included buying commercial food, of course. In a 1935 advertisement for Spratt's biscuits that appeared in both veterinary journals and the popular *Dog World*, an imploring German shepherd asserted that "It's up to YOU to feed me right." The ad sent combined messages: the slogan was meant to inspire guilt in the neglectful (nonbuying) dog owner and provide reassurance for Spratt's customers; the dog was portrayed in a begging pose, yet assertively added that "I am speaking for myself and the doggy rights of all Dogdom."[17] This advertisement simultaneously validated the worth of a companion animal and established guidelines for care that implied negligence if a purchasing standard was not met. By appealing to an owner's sense of stewardship and affection for her or his animal companion, pet food companies helped to create a market demand for their products in the interwar period.

Valuing Affection and "The Great Humane Society"

Most veterinarians, encountering consumers' interest in care for pets, required some time to reframe their attitudes toward animals that had no tangible economic value, but that did have great sentimental value. At the time of World War I, few veterinarians held clients' affection for their pet animals to be as important as the economic value of horses and livestock. Indeed, veterinary culture stressed an antisentimental stance toward the treatment of its patients (see Chapter One). An early and influential pro-

"IT'S UP TO YOU TO FEED ME RIGHT." Advertisements such as this one reinforced standards of pet care by subtly accusing owners who did not purchase the products of mistreating their pets. *Spratt's Patent Ltd. advertisement, from* Journal of the American Veterinary Medical Association 86 *(June 1935): 13;* journal page used with permission.

ponent of companion-animal practice, Kansas City veterinarian J. C. Flynn described most veterinary practitioners in 1913 as "uninterested" in companion animal practice. If a dog owner called the veterinarian for help, Flynn reported, the usual response was "give him a good dose of castor oil and don't feed him for a couple of days." Although they were in need of new patient populations, veterinary practitioners "did not care for dog practice." Moreover, many veterinary schools (especially the remaining private schools) did not include training in companion-animal medicine and surgery before World War I. Flynn maintained that veterinary schools and practitioners should treat companion animals "in the same scientific manner" as other domestic animals, and he ended with a plea to "give the faithful dog the attention he deserves."[18]

Although change was slow, veterinary practitioners did increasingly devote attention and care to companion animals in the 1920s and early 1930s as a result of three factors: changes in the domestic animal economy and the veterinary schools; the influence of a small but devoted group of practitioners (including Flynn) who proselytized on the importance of companion animal practice; and the redefinition of veterinary medicine as a humanitarian (as well as scientific) profession. The search for new patient populations to replace horses did not obviously lead to small-animal pets.[19] However, the most likely candidates—food-producing animals—presented a problem for practitioners during the 1920s because livestock raisers caught in the agricultural depression often could not pay for veterinary care for their animals (see Chapters Two and Three). Meanwhile, pet owners were actively seeking veterinary care. As early as 1913, the "hospital for dumb animals" run by New York City's Society for the Prevention of Cruelty to Animals (SPCA) treated 15,000 dogs and 4,000 cats annually. Especially for those veterinary practitioners who wanted to live in urban or suburban areas, pets represented a rational choice because of the increasing demand. Also, training in companion-animal health was becoming more available (although horses and farm animals were still the main focus of veterinary curricula). Veterinary schools, especially those located in urban areas, offered students increasing amounts of clinical experience with companion animals through their teaching hospitals. By 1923, the University of Pennsylvania's Philadelphia hospital treated 5,937 "small animals"— mostly dogs and cats—while caring for only 838 "large animals"—mostly horses. Cornell, Pennsylvania, and Iowa State all offered courses in canine medicine and surgery in the 1920s.[20]

A small but influential group of practitioners interested in caring for companion animals also provided guidance and encouragement for their colleagues in the 1920s and '30s. J. C. Flynn traveled around the country offering veterinarians educational lectures and demonstrations of his easy spaying (ovariohysterectomy) and castration techniques. New Jersey veterinarian Mark Morris later remembered his colleagues' initial astonishment at his interest in diagnosing the little-explored ailments of companion animals. Morris gave numerous presentations in New York City and at national meetings on his new blood analysis techniques and nutritional formulations for dogs and cats, both developed in a laboratory attached to his animal hospital. Companion-animal practitioners organized the first special section on small animals at the 1925 St. Louis meeting of the American Veterinary Medical Association. All of these public appearances and meetings taught practitioners basic medical and surgical techniques that made them more comfortable with treating the most common companion-animal species and helped lend professional credibility to the enterprise.[21]

However, medical and surgical skills represented only part of the new knowledge that practitioners needed for companion-animal work. Flynn and others also wrote articles for the *Journal of the American Veterinary Medical Association* throughout the 1920s that were designed to introduce practitioners to the philosophical and practical aspects of working with the owners of companion animals. Flynn described guidelines for the veterinarian's personal appearance and the types of assistants he should hire. Flynn stressed the major difference from large-animal practice: "The matter of commercial value is not so often a factor . . . but the matter of sentiment is. You may rope, cast, and hog-tie your large animals and not be criticized by your client . . . but not so with the lady's pet poodle." Flynn's Minnesota-based colleague, A. A. Feist, warned that "times have changed, and so we must change with them." Feist offered an ethnographic study of companion-animal owners for readers of the *Journal* and then sought to explain how these clients would view the veterinarian. He spent much of the article describing the appropriate appearance of the veterinary hospital, veterinarian, and his staff. These authors and others reminded readers that these patients were animals valued sentimentally rather than economically. Nonetheless, they assured readers, sentiment translated into a good business for the veterinarian. "How uninteresting, not to mention unprofitable, small-animal practice would be were it not for the sentiment that goes with it," wrote Des Moines, Iowa, practitioner F. F. Parker.[22]

The introduction of "sentiment" as a basis for patient value, however, caused some serious philosophical problems for veterinarians. Most had been trained within a culture that viewed only animals with economic or utilitarian value as worthy of medical care. Retraining practitioners and refocusing veterinary schools on companion-animal patients presented a challenge for interested veterinary professional leaders and companion-animal practitioners. Fortunately for them, the dictates of the market proved a strong influence in favor of pet practice.

A more difficult conflict of values arose from the growth of sentimentally based pet practice alongside the profession's support for research using dogs, cats, and other companion-animal species as laboratory subjects. While most veterinary practitioners did not actively engage in laboratory investigations, the younger members' education had emphasized research, and their profession's leaders expected all of its members to support it; veterinary medicine had to uphold its credentials as a defender of science.[23] Meanwhile, veterinary leaders were urging practitioners to become ever more reliant on pets (especially dogs and cats) as patients. Pet owners assumed the veterinarian's philosophical commitment to preserving pet animals' lives and well-being, and looked to the profession to validate the emotional ties they had formed with their animals. Acknowledging the use of dogs and cats as appropriate scientific research subjects threatened to devalue veterinarians' patients who were pets. Moreover, practitioners whose patients were pets could ill afford to jeopardize their reputations as compassionate caregivers by appearing to support any cruel use of animals, even in the name of science.

One particular group of citizens interested in animal protection, antivivisectionists, had for decades opposed the use of dogs, specifically, in scientific research. Antivivisectionists repeatedly sponsored state and federal legislation designed to abolish such research. While their efforts were decidedly on the fringe of mainstream American culture, the antivivisectionists had two things in common with companion-animal owners likely to patronize veterinary practices: they valued dogs and cats sentimentally, and they were usually women. The veterinarians who confronted antivivisectionists were invariably men representing a culturally masculine profession whose future depended on not sacrificing its scientific foundations. In the 1920s, the American Anti-Vivisection Society (AAVS) and other antivivisection groups began to target the veterinary profession in legislative efforts.[24] While physicians and allied research scientists had extensive experience in battling animal protectionists, veterinarians were recent (and reluctant) additions to the fray.

Antivivisectionists turned their attention to veterinary scientists because their research programs using dogs and other experimental animals grew tremendously after World War I.[25] The largest laboratories, those of the USDA's Bureau of Animal Industry in Washington, D.C., explored both animal and human diseases and housed experiments on a variety of animal species. Obtaining research subjects from the District of Columbia's dog pound, veterinary researchers began a series of experiments on the diseases of dogs after World War I. They studied the etiology of pellagra (known as "black tongue" in dogs), hookworm infestations in animals and humans, and worked to develop a prophylactic canine rabies vaccine. Numbers of private practitioners, including Mark Morris in New Jersey and David Buckingham in Washington, D.C., also conducted nutritional and toxicological experiments on dogs as contractors for pet food and chemical companies. These projects, although initially small in scope, addressed the concerns of dog owners and their veterinarians in the treatment of canine diseases while often also providing basic research into remedies for similar human pathophysiological conditions.[26] Such experimentation represented a marked change in intellectual focus for veterinarians, who had previously concentrated their research efforts almost exclusively on the diseases of horses and food-producing animals.

Research on dogs had the unforeseen consequence of stimulating antivivisectionists' interest in the BAI beginning in the late 1920s. Many concerned citizens wrote letters directly to the president of the United States challenging the use of dogs in federal laboratories. These were forwarded by the White House to BAI veterinarian Maurice C. Hall, who was designated to answer them. Hall, a parasitologist who conducted the majority of the BAI's dog experiments, became increasingly enmeshed with animal protectionist forces. In 1930, the Vivisection Investigation League introduced what proved to be an unsuccessful congressional bill prohibiting the use of dogs in research in the District of Columbia. During the spring and summer of that year, Maurice Hall represented the veterinary profession at hearings before the House and Senate committees on the District of Columbia.[27]

These hearings, occasionally enlivened by colorful outbursts and protests, tell us much about the shifting ideologies of value placed on dogs and other companion animals by veterinarians and antivivisectionists. Testifying members of the Vivisection Investigation League tried to keep their anti–dog research position clear, but they stumbled when confronted with veterinarians' argument that it was necessary to sacrifice a few dogs to ensure the health of the rest. Unlike physicians (who were the antivivisectionists' usual target), veterinarians could effectively counter antivivisectionists' accusations by asserting that they acted according to the best interests of the canine species. Maurice Hall, for example, testified that "it is a matter of satisfaction to me that I have helped to save the health and lives of millions of dogs annually for years." Hall and his supporters cited his investigations into hookworm and other canine parasitic infections, and his discovery of effective anthelmintic protocols, as evidence of research's humanitarian value for dogs. As David Buckingham testified about Hall's work, "It was [done on] the dog, for the dog." Hall went on to accuse his opponents of hypocrisy: "I am as fond of dogs as any proponent of this bill," Hall asserted, "[and] I have done far more for dogs than any of them have."[28]

Despite the argument that experimentation and animal welfare were consonant goals, antivivisectionists suspected that veterinarians' motives included more than simply a humane concern for the welfare of dogs. They pointed as an example to David Buckingham, who ran a veterinary hospital for small animals and a research kennel on the side. Antivivisectionists believed that Buckingham illustrated their argument that veterinary practitioners would be morally tainted by their involvement with research. As historians have described, physician researchers also fell under suspicion that their passion for science and their indifference to the suffering of laboratory animals would lead to experimentation on their human patients.[29] The animal patients of veterinarians were similarly at risk, argued the Vivisection Investigation League. They accused Buckingham of "conducting a hospital for animals, and they [his patients] are to be experimented on." According to his critics, Buckingham had violated his clients' trust. "Those animals [his patients] are given to him to be cured by people who own them," stated a proponent of the antivivisection bill; the alleged actions of veterinarians supplying animals or research results to laboratories was a "practice which is absolutely despicable . . . not ethical in the least."[30]

David Buckingham and Maurice Hall appeared the next week to defend themselves and their profession, and in doing so articulated two important arguments that defined the nature of veterinarians' duty to their patients and to science. First, Buckingham drew a distinct line between patient and subject, treatment and experiment. He asserted that he had purchased his experimental dogs privately, from an owner who was starving them, and that his experiments had not harmed the animals. Buckingham admitted that like any good practitioner, he often used new remedies on his patients. But these remedies were curative, and always humane: "As a practicing veterinarian, I surely could not produce very much pain, or I would lose what little practice I have, so I have been very cautious." Buckingham further asserted that he carefully separated his experimental animals from the dogs in his hospital. "Those dogs belong to other people. . . . I treat [them] as patients," he stated, and added, "My whole practice is based on sentiment."[31] By creating firm distinctions between pet dogs and laboratory dogs, this type of argument could reassure owners that veterinarians knew the difference in how these animals were valued and would treat them accordingly.

Following the defeat of the antivivisectionists' proposed legislation, Maurice Hall left the hearings determined to straighten out the issue of sentiment versus science publicly. "For various reasons," Hall wrote in 1931, "it has been necessary for me to take an active part in the fight against [antivivisection] legislation."[32] One of those reasons may have been his interest in defending veterinarians' public image. In 1930–31, Hall was serving as president of the American Veterinary Medical Association and thus represented the entire veterinary profession and not merely the BAI. Hall's zeal may also have been spurred by an unpleasant personal encounter with an antivivisectionist at one of the congressional hearings. Two weeks after the conclusion of the hearings, he wrote to a colleague: "As I was leaving the hearings on the second day a woman about 30 years old met me at the door and said: 'You'll get yours when the devil tears you apart in hell!' Her obvious pleasure in contemplating this idea labeled her, more definitely than she could have been aware, as a sadist, and her regard for dogs as a compensatory reaction."[33]

With the image of "the sadistic nature of the proponents of the bill" firmly in his mind, Hall published an article in the *Scientific Monthly* in 1932 designed to combat the "propaganda" of the antivivisectionists.[34] In it, he detailed the veterinary profession's most powerful rhetorical strategy in defense of both animal welfare and animal experimentation. Hall's argument ran like this: the cruelties of man were insignificant compared with the cruelties of nature; the worst cruelty of nature was the suffering and death inflicted by disease; physicians, veterinarians, and other scientists were the experts who protected humans and animals from disease; therefore, these scientists were the greatest arresters of cruelty. In particular, Hall elevated the Bureau of Animal Industry, owing to its work against animal disease, to the status of a "great humane society," greater even than those organizations that protected animals against the comparatively minor cruelties of humans. "No other humane society," he wrote, "can show so many animals protected from human cruelty . . . or from the thousandfold cruelties of nature . . . as can the Bureau of Animal Industry."[35]

By presenting the BAI as a humane society, Hall chose the field of humanitarianism on which to battle with antivivisectionists and caught his opponents in a logical trap. If one accepted the premise of his argument, that disease caused most of animal death and suffering, and if one believed that humanitarian efforts should be directed at reducing suffering and death, then one had to admit the value of combating disease. The applied science of veterinary medicine had successfully combated animal disease because it relied on basic research, which required animal experimentation. Therefore, Hall concluded, animal experimentation was an essential part of the great humanitarian effort on behalf of domestic animals. Hall advocated support for the veterinary profession and its institutions as the proper expert mediators in the fight against "the cruelties of nature." Using this argument of science-as-humanitarianism, Hall presented a rhetorical solution to the veterinary profession's problem of how to avoid alienating pet owners while supporting scientific research on dogs, cats, and other animals. He constructed an identity for veterinary medicine that accommodated both sentiment and science and also took into account the dynamic sociocultural value placed on domestic animals. Veterinary leaders used their AVMA president's argument to their advantage for at least the next 15 years.[36]

Creating Standards of Care

For practitioners, the combination of sentiment and science proved to be a good public relations tool. The veterinarian as sympathetic scientist fit pet owners' expectations well. Veterinarians' concerns with furthering the health of pet patients coincided neatly with their desire to shape consumers' expectations for companion-animal care. Authors of articles in veterinary journals, and their practitioner readers, sought to link people's affection for their animals to the purchase of "scientific medicine." "The men and women who love their pets," explained practitioner J. Elliott Crawford, "will respect your ability in proportion to the care and comfort you give their dogs and cats when they are ill."[37] Veterinarians could create an expectation that pet animals should receive a high standard of medical care by maintaining a competent professional demeanor. "A white gown has a psychological effect on your client," advised practitioner A. A. Feist, "[because] he feels everything is being done scientifically." The scientifically trained veterinarian was prepared to offer the same services for pets that their owners could expect at a hospital for humans.[38] Thus, the veterinarian who treated pets was expected to be both scientific and sentimental. A veterinarian's gentleness and concern for the welfare of dogs and cats brought pet owners to his practice, and their perceptions of his acumen as a medical professional persuaded them to return.

Throughout the depression years, demand for veterinary care for pets continued to motivate more veterinarians to undertake this type of practice. Animals such as cattle and horses plummeted in value, and even thriving large-animal practices declined. Although companion-animal practitioners also suffered, citizens impoverished by the depression amazed veterinarians by "scraping up a few dollars to treat the family pet" at the clinic or hospital.[39] Moreover, veterinarians had few competitors in treating companion animals (unlike the situation with food-producing animals, who could be treated by farmers, animal scientists, and county agents). By the mid-1930s, practitioners located in urban areas had become heavily dependent upon small-animal practice. "In the cities, with the decline of the horse, the profession has gone to the dogs," declared veterinary educator Pierre Fish. "Small animal practice is remunerative and has quite full[y] compensated for the former horse practice."[40] Indeed, both urban and rural veterinary practices included pet animals as patients. Eighty-one percent of veterinarians polled in 1930 saw pets as part of their general practices and were spending more of their professional time on pets than on horses.[41] An increasing number of specialists devoted themselves entirely to pet practice. In the largely agricultural state of Illinois, for example, 150 of the state's 543 veterinarians specialized in companion-animal practice in 1930. Illinois veterinarians concentrated even more on pets between 1933 and 1938: 43 of the 71 men who passed the state's licensing examination during this time period applied for entrance into small-animal practice.[42] Companion animals had clearly become an important and growing market for veterinary practice by the late 1930s.

In the process of maximizing access to this market, veterinarians helped shape consumers' behaviors and expectations for companion-animal health care over the next three decades. They accomplished this by engaging in a variety of activities, two of which particularly stand out: the development of vaccination protocols for dogs and other pet animals; and the creation of appropriate spaces—hospitals—in which to care for them.

As veterinary practitioners developed their new market focus on pets after World War I, researchers and educators explored opportunities to expand their work in new directions. The development of vaccines for companion animals accomplished the goals of both researchers and practitioners while supporting citizens' desire to keep pets. During the 1920s and '30s the interest of veterinary researchers in vaccines for viral diseases fit well into an international program of research on filterable viruses. Veterinary practitioners whose pet-animal practice was increasing needed tools such as vaccines with which to care for their patients. Rabies, a much-feared viral disease, was a particular concern because dogs had long been viewed as the source of human exposure.[43] During the mid-1920s, federal agencies (including the BAI) began to record increasing numbers of confirmed rabies cases in people and dogs. By the summer of 1929, rabies episodes had resulted in the quarantine of twenty-seven midwestern counties over the year. Traditional measures to control rabies included keeping valuable dogs indoors or leashed and killing stray dogs that might carry the disease. However, these measures had proved ineffective to prevent human exposure in the past, and veterinarians and public health officials began to focus on the development of an antirabies vaccine for dogs. "Home dogs," those that belonged to families as valued pets, could then be vaccinated to create a barrier of immunity between people and feral sources of infection.[44]

Although some groups (including antivivisectionists) opposed widespread vaccination, veterinary professional leaders interested in the rabies issue in the 1930s successfully promoted the development of a competent, standardized vaccine and educational campaigns to persuade communities to require owners to vaccinate their dogs.[45] As with tuberculin four decades earlier, veterinarians sought control over rabies vaccination by insisting that state and municipal regulations stipulate vaccination only by a veterinarian. In 1930, two major types of killed-virus vaccine were available in the United States. At the Bureau of Animal Industry, veterinarian Harry W. Schoening and others found the chloroform-killed vaccine to be the safest and most consistently effective. This was good news for Detroit-based pharmaceutical manufacturer Parke, Davis & Company, which was developing and later advertised its chloroform-killed rabies vaccine as a way for veterinarians to "protect [their] professional reputation by using the best."[46] Controversy continued over how often the vaccine should be given (once or twice a year) and its importance alongside the police sanitary measures of quarantine and destroying dogs running at large, but veterinarians settled on the chloroform-killed vaccine as the prescribed formula. Most veterinarians agreed that vaccination should be a part of an appropriate rabies control program and that veterinary practitioners alone possessed the appropriate professional skill to administer it. Veterinarians had long complained that they had been excluded from the group of public health officials charged with combating rabies. This situation changed dramatically with the standardization of the vaccine by the late 1930s and with the efforts of veterinary leaders and practitioners to generate public demand for regulations mandating vaccination.[47]

Massachusetts veterinarians provided an example of how successful a public relations campaign for rabies vaccination could be. A committee of the Massachusetts Veterinary Medical Association (MVMA) wrote and printed ten thousand copies of a pamphlet on rabies prevention. MVMA members distributed the pamphlets to clients, and local boards of health requested copies to be given to dog owners when they licensed their animals. MVMA practitioners then offered their services as vaccinators at public clinics. Municipal boards of health employed them (at low cost) to run almost seventy rabies vaccination clinics around the state in 1936. Of 40,000 licensed dogs in Massachusetts, 21,000 were vaccinated at these public clinics (with still more vaccinated during visits to veterinarians' offices). Each vaccination represented an opportunity to educate dog owners about the importance of annual vaccination and veterinary examination. Dissemination of such information helped to combat the claims of antivivisectionists, kennel owners, and other critics that the disease did not exist or that vaccination was merely a "racket" designed to enrich veterinarians. For their part, veterinarians were willing to provide physical examinations and vaccinations for pets at occasional public clinics, at lower cost than a private office visit, as a way of creating clients who would return year after year for full-fee revaccination and examination. Beyond this, the key goal was to get "public opinion to demand enforcement of control measures"— including canine antirabies vaccination. By running public clinics, printing pamphlets, and publishing articles in newspapers and magazines, veterinarians sought to convince municipal and state health boards and dog owners that annual rabies vaccination and examination by a veterinarian was a part of proper dog care and a necessary public health measure.[48]

Distemper (another viral disease) concerned animal owners because it was a major killer of dogs in the early twentieth century. The news that an English team of researchers had developed a prophylactic serum in the late 1920s galvanized veterinarians in the United States. "No discovery of greater importance to dog-owners and veterinarians has ever been announced," editorialized the *Journal of the American Veterinary Medical Association* enthusiastically, "and there is already a keen desire on the part of many veterinarians to obtain this new immunizing treatment."[49] Small-animal practitioners reproduced and tested the prophylactics themselves, some independently and others under contract with pharmaceutical companies. Practitioner-experimenters such as Mark Morris and George Watson Little communicated their results to their colleagues at meetings and in the veterinary journals.[50] Indianapolis-based Pitman-Moore was the first company to receive a federal license to produce and market a distemper vaccine. By 1935, Lederle, Pitman-Moore, and Jen-Sal each offered three or more types of distemper prophylactics for veterinarians to choose from and use as an annual vaccine in their practices. This variety reflected the fact that controversy about the nature of distemper itself and its prevention continued among scientists through the 1930s. Some veterinarians believed that distemper was caused primarily by bacteria rather than filterable viruses; virus proponents countered that the bacterial infections were merely secondary.[51] Regardless of which product they chose to use, practitioners and pharmaceutical companies worked in close partnership: advertisements in veterinary journals made it clear that such vaccines were for sale to graduate veterinarians only.[52]

Cats were also prone to viral diseases, and owners of catteries and individual pet animals sought vaccination for their animals. Although they were less popular than dogs at first, cats grew in importance as commercialized pets after World War II. Feline pets fit well into apartments and small houses and were reputed to make fewer demands on their owners for attention and care. A popular press devoted to cats expanded greatly after

World War II. *Cats Magazine* was founded in 1945, and popular advice manuals such as *A Practical Cat Book, The Complete Book of Cat Care,* and *The Fabulous Feline* came out after the war, to be followed by many other publications on cats. By 1958, cat food sales were growing faster than those of dog food, and cat food accounted for 26 percent of all pet food sold. Pharmaceutical companies had also taken note of the growing interest in cats. As early as 1935, Lederle Laboratories offered a feline vaccine consisting of killed enteritis and distemper viruses for administration by veterinarians. Cats too deserved "preventive medicine," the shorthand term veterinarians used to describe the annual vaccination and physical examination at the veterinarian's office.[53]

Preventive medicine for companion animals simultaneously received the blessing of public health authorities, brought veterinary practices a steady flow of income, and reinforced standards of care for pet owners. Veterinary practitioners and pharmaceutical companies cooperated to invent methods of reminding pet keepers about their responsibility to have their animals examined and vaccinated annually. Pitman-Moore published pamphlets designed to educate dog owners on distemper and rabies. Not coincidentally, the company promised that the pamphlets would help the veterinarian increase his practice; practitioners purchased them cheaply and distributed them to their clients. By the late 1930s, veterinarians were also using "reminder cards," which they found to be a successful method of reminding owners to return annually for their pet's vaccinations and physical examination. The reminder card could be placed in a monthly file for the next year and mailed to the client at little expense to the veterinarian. In this way, veterinarians again communicated the importance of repeated examinations and prompted pet owners to act "responsibly."[54]

By the mid-1950s, veterinarians had made pet vaccination a cornerstone of their position as public health experts and of their rapidly growing companion-animal practices. Campaigns for rabies vaccination had been conducted as a public health measure, but distemper vaccination was strictly a dog health measure. Although only rabies vaccination was mandated by public health officials (especially after Rockefeller researchers developed an improved vaccine in 1950), both rabies and distemper vaccines were administered during the dog's annual office visit. Veterinarians from California to Michigan to New York, through their local veterinary associations, continued to conduct periodic public vaccination clinics well into the 1950s. The public relations advantages of holding clinics evidently outweighed the serious criticisms of them from within the profession. Most practitioners must have agreed with Michigan veterinarian D. J. Francisco, who wrote that he and his colleagues did not "favor the public clinics with reduced vaccination fees and free veterinary service as we feel it approaches socialized medicine."[55] Despite these objections, Francisco himself headed a successful public clinic in his home county in 1954. Practitioners' willingness to conduct public clinics devolved from their interest in furthering public health measures and in establishing a patient population of pet animals who required preventive as well as acute care. While many pet owners would surely have avoided vaccinating their animals, municipal, county, and state regulations maintained surveillance of the dog population and tightened control on licensing and rabies vaccination, especially in cities. Moreover, as *Field and Stream* editor Joseph Stetson reminded pet and working-dog owners, "the dog population stands to gain more from vaccination than the human population." In this view, an interest in pet-animal welfare necessarily included vaccination. Urged by veterinarians, public health officials, popular magazines, and news articles to get their dogs vaccinated, pet owners around the nation increasingly (if reluctantly) complied. By the 1960s the annual physical examination and vaccination for rabies and distemper had become standard care for pet dogs in much of the United States.[56]

Veterinarians' promotion of the annual visit for vaccination and examination paralleled their creation of new facilities appropriate for companion-animal care. Along with vaccination and other preventive care, animals periodically needed hospitalization for illness or injuries. With notable exceptions, small animals in veterinary hospitals had traditionally been relegated to isolated cages in the "dark corners" of box stalls meant for equine patients. This would not suffice any longer, warned small-animal specialists. Pet owners were savvy consumers, and they expected their animals to be cared for in clean, comfortable surroundings that resembled a human hospital rather than a stable.[57] An animal's owner "leaves his pride and joy to our tender mercy to be cared for," explained J. Elliott Crawford in 1929. Such an owner's "confidence" had to be earned, and she or he was more likely to patronize a veterinarian in whose "institution he takes a pride." Indeed, wrote Los Angeles practitioner W. E. Frink the next year, veterinarians would be "surprised at the extra monetary return besides the satisfaction in having a suitable place to care for such [small-animal] cases as occasions demand." A veterinarian's hospital or clinic was an important attraction to customers.[58]

A hospital's design had to reflect the veterinarian's "intent and purpose," which included what W. E. Frink called "scientific and humane" (as well as economic) considerations. The animal patient deserved a comfortable bed, cleanliness, suitable food, and proper medical and surgical care while in the hospital. The animal's owner should be allowed to see the ward where his or her pet would be kept. Owners, Frink asserted in 1930, would be willing to pay for excellent accommodations and service. He suggested a charge of $1 per day for hospitalization and an additional daily dollar for the veterinarian's medical care (not an insignificant sum during the depression). Texas practitioner Horst Schreck applied principles similar to Frink's in his 1931 series of articles on hospital and clinic design in the *Journal of the American Veterinary Medical Association*. Schreck described in detail the proper design of sanitary but comfortable kennels, efficient but homey reception rooms, and state-of-the-art operating theaters and examination rooms. All of these attributes could be had, he argued, in a building that had formerly been a horse stable, or one that included living quarters for the veterinarian and his family on the second floor. Above all, hospitals and clinics should be designed to increase the veterinarian's efficiency, magnify his medical and surgical skills, and persuade the "animal-owner that we think quite as much of his animal as he does."[59]

These writings and others underscored the necessity for veterinary practitioners to carefully consider the services they offered pet owners in an expanding market. In his article, Frink asserted that "the future of the small-animal hospital depends upon how we respond to the rapidly growing demand for the services of such an institution." Veterinarians could shape their own destiny, he implied, by encouraging pet owners to continue seeking medical care for their prized animal companions. Because "the modern owner of a dog or cat has had little or no training in animal industry," veterinarians supplied crucial services that would reward them with "the confidence and respect of the public."[60] Renovating or building a hospital was an important component of practitioners' efforts not only to meet pet owners' needs, but also to shape their expectations. The demand for reprints of Horst Schreck's hospital design articles was so great that the AVMA published them as a booklet and sold it to practitioners throughout the 1930s. Schreck published further hospital design articles in *Veterinary Medicine* that were again reprinted in 1946. With the availability of building materials and the general prosperity of the postwar years, practitioners' demand for the articles exceeded the journal's ability to print and mail them.[61]

Veterinary hospital building and small-animal practice in general also benefited from the efforts of the American Animal Hospital Association (AAHA), especially in the 1950s and 1960s. Founded in 1933 by J. C. Flynn, Mark Morris, and other pioneer small-animal practitioners, the AAHA served as an exclusive specialty organization. Selective in its admission standards, the AAHA grew slowly in influence until the early 1960s. Its goals included regulating small-animal hospital facilities (AAHA accreditation was a notoriously difficult process) and establishing standards for small-animal care that were based on the American College of Surgeons' regulations for human hospitals. As the number of exclusively small-animal practices grew (and other types decreased), the AAHA's influence in the veterinary profession also grew. Along with the AVMA, its standards and publicity efforts helped city and town officials to view veterinary hospitals as "professional facilities that deserve special attention" and as "acceptable features of any community." AAHA accreditation also served to reassure animal owners that they were purchasing the best medical care available for their pet animals.[62]

By the 1960s, more owners were demanding specialized hospital services for cats as well as for dogs. Most veterinarians had traditionally been uninterested in cats because few owners had sought feline medical care and veterinary schools taught little about cats. In the late 1960s and early '70s, however, the growth of the pet cat population and demands of cat owners stimulated several professional changes. Veterinary schools added information on cats to their curricula and pharmaceutical companies increased their range of products available for cats. Interested practitioners founded a professional journal on feline health in 1971 and the American Association of Feline Practitioners (AAFP) in 1974. The AAFP assessed practitioners' existing protocols for vaccination, examination, and treatment and provided standardized recommendations for cat care. Much as dog proponents had argued for the canine right to appropriate medical care in the 1930s, practitioners interested in cats insisted that these animals deserved specialized treatment. Barbara Stein, a pioneer feline practitioner, later remembered that "I didn't understand why a ten-pound Persian might be valued less than a ten-pound poodle." The veterinary profession's new feline focus had arisen from cat owners' "concern" for their pets, translated into their willingness to purchase health care for their animals.[63]

With the establishment of modern animal hospitals and vaccination and treatment protocols between 1935 and 1970, veterinary researchers and practitioners defined the health-care services that consumers could purchase. The availability of these services set standards for appropriate animal care that could be cited to reprimand pet owners who were unwilling or unable to purchase these services. By raising the standard of companion-animal care, veterinarians helped to create the "priceless" pet—an animal whose owners' affection for it dictated that they spend more money on its care than the creature's economic worth. By the end of the twentieth century, priceless pets were well established in American society and popular culture. Veterinary professional leaders, researchers, and practitioners working to provide preventive and hospital services for pets had created a consumer imperative that included the use of technologies (vaccines), specialized facilities, and the proscription for pet owners to include veterinary services in the care of their animals.

With their new technologies and hospitals, veterinarians felt they deserved to be recognized as the experts on companion-animals' health issues. They had earned a position of public trust and authority comparable to that of their medical colleagues. "If you were an ailing dog," explained an article in the *Science News Letter* in 1951, "you'd probably be treated with essentially the same techniques that medical doctors use on human patients." Only *some* ailing animals received such treatment, of course; the entitlement to medical care depended on the animal owner's valuing her or his pet enough to purchase it. But for priceless pets—those who had attained a humanlike position in the family—the same type of imperative for care that historians of medicine have identified for human patients applied to animals as well.[64] If a pet's health required a particular procedure or technology, then many owners felt morally bound to seek it. This belief was reinforced by veterinarians and other authorities who asserted that the results—increased animal health and welfare—justified the owners' willingness to purchase medical care. J. C. Furnas told *Saturday Evening Post* readers in 1955 that "modern medicine has raised the American dog's standard of living," citing estimates of dogs' longer life spans, decreases in diseases such as distemper, and a new antihepatitis virus vaccine soon to be integrated into the annual distemper shot. Furnas cited "the increasing public consciousness of what vets are now set up to do" and owners' interest in the best treatment for their pets. As a subsequent *Saturday Evening Post* article declared, the veterinarian was an essential expert in times of trouble—"your dog's next-best friend."[65]

By the mid-1970s, one *Journal of the American Veterinary Medical Associa-*

C.Barsotti

"The bidding will start at eleven million dollars."

THE PRICELESS PET. As immortalized in this 1985 *New Yorker* cartoon, a companion animal's price reflected the value of its owners' affection for it. *Copyright The New Yorker Collection 1985, Charles Barsotti, from cartoonbank.com.*

---

tion author evaluating the growth of pet practice asked, "Is there an endpoint to the fees that the public is willing to pay to cover the increasing cost of quality veterinary care of pet animals?" Of course, his question applied only to those animals fortunate enough to be "members of the family." Many others of the same species suffered impoundment, homelessness, cruel treatment, life as a laboratory subject, and death. Veterinarians recognized the differences in value applied to individual animals of the same species. As Maurice Hall had explained the differentiation between pet dogs and laboratory dogs, "having a home gives assurance of a dog's economic status . . . society agrees that dogs [may] be killed for an unsound economic and social status." Attaching "social status" to individuals was crucial to veterinarians' objective of encouraging pet owners to purchase health care for their animals. This is not to deny that practitioners had compassion for animals that did not qualify as priceless pets. Veterinarians worked for humane organizations and treated charity cases, and their hospitals often sheltered unwanted animals. However, companion-animal practice had remained remunerative because of veterinarians' efforts to validate the special status of priceless pets and to establish a standard of health care for them.[66]

Companion-animal practice also helped to redefine the veterinary profession's mission and culture in the mid-twentieth century. From its early days as a profession in the 1890s, veterinary medicine's leaders envisioned their mission as one of protecting the health of animals valuable for their contributions to the national economy and public health. While the means to these ends frequently invited controversy and dissent, few if any of those calling themselves "veterinarians" diverged from the profession's stated goals. The culture of veterinary medicine, emerging from the barnyard and livery stable as well as the laboratory, dictated a focus on those animals that provided food and transportation, thus supporting the mission of the profession. However, as pets came to have a greater importance for many owners, the change partially restructured the value system applied to animals by translating emotional attachment into otherwise inexplicable economic value. This reality forced veterinary leaders to rearticulate the profession's identity as a "great humane society," a definition crafted to be congenial with the dictates of animal welfare, public health, and market economics.

Veterinarians' increasing focus on companion animals precipitated important changes in the attributes of their clients as well. In particular, urban and suburban women joined and eventually outnumbered the livestock and gun dog owners that veterinarians had been accustomed to serving. Companion-animal practitioners had long been aware of the changes that the new client demographic demanded. A. A. Feist warned veterinarians in 1930 that clients, many of whom were "ladies and children," expected the veterinarian and his staff to be animal lovers. The presence of a woman on the clinic staff was "the one great secret" to making this positive impression, he argued.[67] The female staff member was, of course, the receptionist or bookkeeper, and often also the veterinarian's wife.[68]

Although companion-animal practice was seen as a more "fit occupation" for a female veterinarian than livestock work, the expansion of pet practice made little immediate difference in the tiny numbers of women allowed to enroll in veterinary schools in the United States. However, a higher proportion of female veterinarians than males did practice companion-animal medicine from the 1930s onward. The first woman to hold an office in the AVMA, Helen Richt Irwin, served as the secretary of the section on small animals in 1937. Women finally joined the profession in appreciable numbers in the early 1970s, owing in part to social interest in gender equity manifested in federal regulations. The reluctance of veterinary schools to admit women threatened them with a loss of federal funding; revised admissions policies meant that by the mid-1980s, women made up 50 percent of entering classes.[69]

Since the early 1970s, both male and female veterinarians have increasingly specialized in companion-animal practice. Many of the veterinary medical and surgical subspecialties and most veterinary researchers concentrate largely on the health problems of companion animals. Along with the country's population since World War II, veterinarians have located their practices increasingly in urban and suburban rather than rural areas. Many have had little experience with livestock, and most wish to return to the type of urban-suburban areas from which they came. With small-livestock operations in severe decline, veterinarians interested in farm animals have had to compete for a diminishing number of specialty jobs and herd-health consultancies. Meanwhile, the scope of companion-animal work has continued to expand steadily, even within general practices.

To the dismay of some veterinarians, companion-animal practice has become the image that most Americans now associate with veterinary medicine. Yet the profession remains ideologically diversified; it would be unthinkable to abandon traditional interests in public health and the livestock economy. Instead, veterinarians have continually redefined themselves to accommodate the changing values applied to animals without sacrificing long-standing professional goals. They have also sought to mold Americans' behaviors and expectations with regard to their animals. As companion animals became "priceless" in the twentieth century, veterinarians found ways to translate pet owners' regard for their animals into specialized care—and expanded their role as mediators of Americans' relationships with their domestic animals.

CHAPTER SIX

Reconciling Use and
Humanitarianism

The word *humane* exhibited one of its many meanings when animal protectionists first used it as a title for their umbrella organization, the American Humane Association, in 1877. Etymologically, this word was properly concerned with the welfare of humans, not animals, but the association's organizers held that the principles of charity, compassion, and kindness could be equally applied to either.[1] Arguments for animal welfare often implied that such actions also morally benefited humankind, in both individual and social forms. State bureaus of child welfare arose from societies for the protection of animals. Moreover, an individual's propensity to mistreat animals had long been correlated with hostility to other humans. The opposite was also assumed true; a child's proper moral education had included kindness and charity toward domesticated animals since the eighteenth century in Anglo-American culture.[2] Accordingly, by the beginning of the twentieth century, the idea was well established in the United States that no civilized society condoned unbridled barbarism toward living creatures because such behavior violated social morality and threatened refined modern sensibilities.[3] However, this tenet of civilized behavior, which was rhetorically useful, certainly did not describe the reality of many domesticated animals' lives. The fact that humans put some animals to work and killed others for food created opportunities for some very uncivilized behavior. This was especially true in the United States in the twentieth century, when burgeoning urban populations and a reliance on virile industrialism required human consumption of animal resources on an unprecedented scale.

Indeed, the use of creatures for human needs and desires without sacrificing basic ideals of humane treatment was a central dilemma for people who were in regular contact with living animals. Maintaining moral refinement and meeting the challenges of modernity certainly required both consideration of animal welfare and a vast increase in animal agriculture. Literary analyst Marian Scholtmeijer has described this predicament as a "dissonance" between the "civilized ideals" maintained by an urbanizing society and the ill treatment of animals.[4] In addressing this social quandary, veterinary researchers and practitioners helped to create a value system that validated Americans' needs and desires for animal resources, while maintaining a representation of American culture as one of civilized kindness toward living creatures.

The beginning of the twentieth century was a crucial time for the reconciliation of humanitarianism with the demand for inexpensive and plentiful animal resources. The maturation of the United States as an industrial power needs little explanation here. The population had already begun the tremendous expansion that would continue throughout the century, with most of the increase located in urban areas. In response to the demands of these urban residents, thousands of horses expended their lives in cities and towns voracious for transport and power in the 1900s, '10s, and '20s. Increasingly, the problem of providing affordable food determined the practices of livestock husbandry and slaughter, and greatly expanded this sector of the workforce. As census analyst Harry McCarty wrote in 1902, "the process of converting live stock into food for human consumption is an industry that, directly and indirectly, furnishes employment to a considerable portion of the population of the United States, and sustenance to all."[5] Industrialization, too, devoured ever-increasing numbers of animals whose body parts ended up in books, hats, furniture, and numerous other articles useful or necessary to ordinary life. Industries and other occupations associated with animals employed one in fourteen American workers in 1900. Thus, animals played central roles as the raw materials necessary for food, consumer goods, jobs, and profits at a time when population and consumerism were expanding to unprecedented levels.[6]

At the same time, ideas of humanitarianism toward animals had become firmly established in American culture. Law, medicine, and theology had all included concern for animals as a cultural belief and social guideline in the early years of the country. Thomas Paine in *The Age of Reason* asserted that "everything of cruelty to animals is a violation of moral duty," while in November 1807, the eminent Philadelphia physician, Benjamin Rush, delivered a course of lectures to his students on "the duty and advantages of studying the diseases of domestic animals." Citing the biblical injunction to stewardship, Rush asserted that domestic animals were also owed consideration in return for their contributions to human society and for the ills they suffered under domestication. Rush also speculated on the nature of animals and their relation to humankind: "As moral evil and death accompanied each other in the human race, they are probably connected in the brute creation," asserted Rush. "[Animals have] a probable relation to us in a resurrection after death, and existence in a future state." By asserting that animals could claim not only sentience and cognition, but perhaps also *souls*, Rush situated them disconcertingly close to human beings—and the concept was especially chilling given the often severe treatment animals received. God's judgment of humans' failure to observe their duty to fellow creatures could be harsh, Rush implied (although in the next breath he contrived a justification for eating meat). Nonetheless, ideas such as these contributed to the passage of laws against animal cruelty as early as 1829 in New York State and 1834 in Massachusetts.[7]

The arguments Rush used were familiar to those concerned with animals a century later, although the specific disputes assumed different forms. Veterinary leader Veranus A. Moore followed Rush in citing the biblical statement that humans bore responsibility for protecting the well-being of animals.[8] Moore defined the veterinary profession in part as one "developed to relieve suffering."[9] He equated veterinarians' ministrations with animal owners' interest in the welfare of their creatures. Moore asserted that "veterinarians are servants of animal owners and of the public; they can serve directly only when called upon to do so."[10] Thus acknowledging the difficulties of balancing humane treatment with more material concerns, Moore extended the burden and blessing of using domestic animals, and the need to address the accompanying moral claims, to all citizens, not just Benjamin Rush's "animal physicians."

Animal humanitarianism fit within a larger set of concerns about injustice and the social conscience during the Progressive Era at the beginning of the twentieth century. Francis Rowley, president of the Massachusetts Society for the Prevention of Cruelty to Animals, identified the efforts of his organization with child saving, slaughterhouse reform, and wholesome food, all of which concerned "women who may be counted upon to champion almost every righteous cause."[11] Thus reformer Caroline Bartlett

Crane's interest in meat inspection could not fail to advance the cause of humanitarianism even as it was concerned with maintaining meat eating and the necessarily harsh process of food-animal slaughter. Linking all of these reformist efforts reinforced the "the humane idea" as a sort of moral umbrella protecting Americans from the storm of industrial modernity, "to widen still further the spirit that unceasingly pleads for the just and kindly treatment of all sentient life . . . in our modern day."[12]

This interest in preserving moral integrity met its opponent in the practices required to secure maximal profits; and in the cases of many working and food-producing animals, the latter dictated the treatment they received. Horses beaten in the streets, cows that had collapsed from illness, hogs and chickens transported and slaughtered ruthlessly—these were all common sights that horrified onlookers. The public nature of these spectacles only magnified the distress of people who believed that such behavior violated the moral principle of humanity as well as injuring its victims. Symbolic of other problems, feeding animals into the cruel maw of commercial production challenged the survival of humanitarianism—the very thing a modern society, with its increasing distance from the positive moral influence of the natural world, could least afford to lose.[13]

Guiding the Humane Use of Animals: Veterinarians' Role

Throughout the twentieth century, veterinary researchers and practitioners adopted various programs that addressed the problem of reconciling animal use and humanitarianism. They did so by balancing the interests of those who valued animals in various ways, while building their own professional authority. They helped create and responded to a system of valuing animals that embraced both profit and humanitarianism. Few if any of these programs proceeded without a good deal of resistance from animal producers and consumers; indeed, the examples in the preceding chapters have highlighted the complexity of ideological and practical alliances and disagreements surrounding the negotiation of animal value.

First, veterinarians contributed toward reconciling use and humanitarianism by helping to remove food-animal production and processing from most Americans' daily experience. They did this in three ways: by reinforcing the equation of good animal care with healthy milk and meat products; by functioning as the tools of state assurances of wholesome food that made animal production procedures become invisible to most consumers;

and by finding ways to control diseases that arose from confinement production practices. Invisible production of animal-derived foods was crucial
to reconciling humanitarianism and use of these animals, since few people
could view procedures such as animal slaughter as benign processes, no
matter how well conducted they were (and even fewer people were willing
to stop ingesting meat and milk). Veterinarians' actions encouraged belief
in the idea that the welfare of animals had to be considered for them to be
healthy and produce healthful food. As the Metropolitan Life Insurance
Company reminded its subscribers in 1929, gold-standard certified milk
came only from cows "kept as clean as race horses," in ideal stables and pastures, and examined and tested for tuberculosis by a veterinarian. The Carnation Company carried this theme of animal happiness equals healthy
food even further in its advertising campaign throughout most of the 1900s
by asserting that its products came from "contented cows." Americans
could feel good purchasing products that implied humanitarianism toward
animals while linking this concern to wholesomeness.[14]

As experts and representatives of the state, veterinarians provided the
assurance (however fragile) that animals used for food were healthy and
that the resulting food was wholesome and safe to eat. "Certified milk,"
"U.S. Inspected"—these and other stamps of approval on food meant symbolically that state authority (usually through the office of veterinary inspection) had taken care of issues of animal health and contamination by
disease-causing organisms. As animals such as milk cows and chickens left
backyards and moved into distant production units, stamps of approval
replaced the scrutiny of consumer watchdogs such as Caroline Crane. Historian Jacqueline Wolf has found that at the end of the 1920s, Chicagoans
"enjoyed the luxury of taking clean cow's milk for granted."[15] By the 1950s,
poultry producers were well aware that the "sanitary" condition of meat
had fallen behind price, aesthetic appeal, and convenience as a criterion for
housewives' choice of purchases; consumers simply did not have to worry
about meat sanitation anymore.[16] With the healthfulness of food (and thus
welfare of the animals) ensured, most consumers valued food-producing
animals primarily for the low cost of the products of their bodies.

This low cost, of course, depended on the development of highly intensive methods of raising food-producing animals, and veterinarians contributed to this too. As we have seen, infectious diseases ran quickly through
large groups of closely confined animals, killing them or decreasing their
productive capacities. Veterinarians provided management plans, care dur-

ing epidemics, antibiotic regimens, and vaccines that safeguarded confined populations of animals. These procedures protected the profits of animal owners and kept consumers' prices low. In the bargain, consumers acquired even more distance between themselves and animals that were increasingly housed behind the impermeable walls of huge buildings located in isolated rural places. The post–World War II practice of raising animals in confinement has made the lives and deaths of food animals completely invisible to most Americans. And, as historian William Cronon has put it, the separation of production and consumption that developed over the twentieth century had "moral as well as material implications." The deaths of billions of chickens, cattle, hogs, and sheep could be "unremembered" by marketing the attractively displayed packages in the supermarket as wholesome products from healthy (thus happy) animals.[17] Food companies encouraged consumers to deny that their dependence on animals challenged their humanity; they needed to know little to nothing about animals' actual conditions of living and dying to enjoy purchasing animal products. Having ceded responsibility for humanitarianism to animal producers, veterinarians, and others directly concerned with animal raising, consumers did not need to think about the moral implications associated with using food-producing animals.

This is not to say that the notion of humanitarianism receded from Americans' relationships with domesticated animals in the decades following World War II. Indeed, humanitarianism formed an important foundation for the widespread commercialization of human-animal companionship—the twentieth-century form of pet keeping in the United States. Veterinarians supported the development of commercialized pet keeping by deciding to provide medical services for animals that were valuable solely because of their owners' affection for them. In the second half of the 1900s, most Americans who desired daily contact with animals focused on pets (usually dogs and cats). Pet animals fit very well into a number of existing cultural and social patterns. As Yi-Fu Tuan has described, humans could maintain dominance while simultaneously conducting an affectionate relationship with their animals. Harriet Ritvo's finding that human relationships with animals mirrored Victorian social hierarchies also applied to twentieth-century America's economically driven society.[18]

For twentieth-century Americans, human relationships with companion animals also offered an attractive solution to the central problem of reconciling humanitarianism with use. Americans with pets could treat the animals under their direct control well and consider themselves to be humanitarians toward *all* animals. Treating pets well meant, among other things, purchasing the goods, services, and expertise that veterinarians offered. Veterinarians, accustomed to limitations on the value of working horses and food-producing animals, found that affection commanded prices far exceeding an animal's market value—and that pet owners were willing to pay. One compelling explanation for this priceless affection was an owner's interest in being a self-styled moral person, in maintaining proper stewardship of a dependent creature. This exercise of humanitarianism did not merely reflect simple displacement that was due to the guilt of using food-producing animals.[19] Along with this consideration, companion animals were attractive as packages of many good things: they were representatives of nature, social status symbols, proper teachers of children, loyal friends, and much more. These good things all referred to the necessity of pets' largely middle-class owners behaving according to an acceptable moral standard of humanitarianism. Companion animals made a responsible pet owner's humanitarianism self-evident even as she purchased food and clothing that derived from the deaths of other domesticated creatures.

Humanitarianism and use also could be negotiated by reinforcing distinctions of animal value and creating standards of animal care and use—a role that veterinarians performed in various capacities throughout the twentieth century. These activities in no small part shaped and reshaped veterinary medicine as a profession during the 1900s, as changes in animal value forced veterinarians to create new visions of their professional purpose. At the turn of the century, veterinarians concentrated on the most valuable animals in the United States, working horses, and lived in large cities where the density of their patients was greatest.[20] Horse owners understood their animals' value as a function of their importance to business. Equine illness meant lost business, which translated into lost profits. Veterinarians could thus count on the value of horses, through their work, to ensure that owners would seek and pay for medical care.

By providing medical care, veterinarians helped to reconcile the nation's use of horses with the ideals of humanitarianism. Certainly, the daily work of horses in the country's large cities left much to be desired, according to most accounts. Horses worked outdoors in all weather. They were left to the devices of careless or cruel drivers, and often were denied time off, clean stables, or wholesome food. All of these conditions and more concerned

Americans seeing them and reading about them in everything from the reform-minded *Charities* to the entertaining *Saturday Evening Post*. While acknowledging these conditions, veterinarians proposed the remedy: good business practice as well as good humanitarian practice demanded good care—especially medical care—for working horses. Veterinarians defined themselves early as a profession that had an animal welfare–oriented mandate based in part on criticism over the ill treatment of valuable horses.[21]

Based on this early vision of themselves as a profession, veterinarians thus reinforced the necessity of use and commercialism, linked directly to humanitarian concern, as arbiter of an animal's worth. Given this, the idea that animals valued solely for their companionship could also represent an opportunity for an increased scope of professional authority and patient population came slowly to veterinarians. However, once they decided to promote the idea that affection could be encoded in an economic relationship, they worked to create the mechanisms by which dogs, cats, and other animals became entitled to medical care. Veterinary researchers and practitioners developed specialized animal hospitals, vaccination protocols, and guidelines for care that created the standard for humanitarianism toward pet animals. As an early-to-mid-twentieth-century invention, this standard of caring translated directly into the purchase of goods and services designed to take advantage of consumers' desire to display their increased wealth and leisure.[22]

Veterinarians did more than provide services, however; they also helped to create and codify distinctions among uses of animals, even animals of the same species, that were valued differently. As the prototypical pet, dogs were often the subjects of medical attention, kindness, and careful husbandry; yet they were also abandoned to die on the streets, in municipal shelters, and in laboratories. The veterinary profession, self-proclaimed advocates of better lives for pet dogs but also users of dogs as research animals, taught its members to distinguish between socially valuable and worthless animals and to act accordingly. As BAI veterinarian Maurice Hall asserted in 1930, "having a home gives assurance of a dog's economic status. . . . society agrees that dogs [may] be killed for an unsound economic and social status."[23] By thus describing and reinforcing the distinctions in Americans' valuation of animals, veterinarians reconciled humanitarian concerns with use and supported a system that allowed different valuations for different animals.

Finally, veterinarians also supported this value system by working to marginalize or otherwise contend with dissenters that threatened to upset the social balance of humanitarianism and use of animals. Their response to the antivivisection movement in the early decades of the 1900s provides a case in point. Antivivisectionists had departed from more mainstream animal protection groups in the nineteenth century because, as Harriet Ritvo has put it, "anti-vivisectionists understood the meaning of cruelty differently from the humanitarians who dominated the RSPCA [Royal Society for the Prevention of Cruelty to Animals]."[24] Antivivisectionists in the United States likewise disagreed with American humane organizations, implicating the harsh materialism of modernity in their opposition to the use of animals in scientific research. Thus, veterinarians' opposition to antivivisectionists did not jeopardize their alliances with more mainstream humane organizations.

American veterinarians contributed directly to the defense against antivivisectionists for two reasons: they needed to remain in good standing with the scientific community, and they wanted to police the system of animal value that they had helped to create and validate. Thus, when the BAI's John Mohler provided W. W. Keen with specifics of animal anatomy, physiology, and husbandry throughout the 1920s for Keen's proresearch lectures, he ensured that veterinarians would be listed alongside physicians as scientists and defenders of medical research.[25] Veterinarians also testified against antivivisectionists at congressional hearings over proposed restrictions on the use of animals in research and drummed up support among practitioners. While most veterinary practitioners did not actively engage in laboratory research, the younger members' education had emphasized it, and their profession's leaders expected its members to support it. Practitioners were willing to do so as long as their pro–animal welfare reputation, and the economic viability of their practices, were not damaged.

By reassuring pet owners that research benefited all animals, and reinforcing the different values applied to pet animals and laboratory animals, veterinary leaders linked sentiment with science and supported practitioners. Their arguments worked so well because they reconciled concerns about welfare with the need to use animals as laboratory subjects. The Bureau of Animal Industry, and particularly veterinarian Maurice Hall, functioned as the government's source of response to antivivisectionists in the first decades of the twentieth century. Hall set the tone for the veterinary profession at large by replying to letters written by concerned citizens and by instructing veterinarians how to reply to similar local inquiries. Hall identified veterinary medical research as a humanitarian effort, since its purpose was to prevent the suffering that accompanied disease and disability in animals.[26] Of course, it also prevented loss of profits in the livestock industry, a fact that made Hall's arguments quite functional for the bureau. Furthermore, veterinarians such as Hall pointed to the profession's relationship with American humane organizations as evidence of its humanitarianism. This relationship served a further purpose: it reinforced veterinarians' scientific research agenda as an appropriate component of a compassionate profession in a society that sought to balance kindness with more practical concerns.

Veterinarians opposed antivivisectionists out of professional and social interest. Antivivisectionists based their goal of prohibiting all use of animals in research on a moral argument that included humanitarian and animal rights components. By using the concept of animal rights to challenge the widely accepted use of research animals, antivivisectionists placed themselves in conflict with the system of animal value endorsed by veterinarians. If animals had moral rights to identity and self-determination, then every use of an animal for human purposes violated morality. Thus, veterinarians' rejection of the antivivisectionist position not only helped to secure their position as scientists but also protected Americans' practice of assigning values to animals so that they could be used and benefited from with no loss of moral standing.

Veterinarians (like physicians) worked against antivivisectionists by pitting the rational masculinity of their proresearch position against what they characterized as the strident and unreasonable demands of opponents to animal research. By the beginning of the twentieth century, it was not tremendously difficult to marginalize antivivisectionists. Their movement had reached its social zenith in the nineteenth century; its arguments remained inflexible though times had changed; and its proponents (although active and determined) were largely middle-class women (rather than upper-class men).[27] Veterinary leaders focused on this last characteristic in their efforts to discredit antivivisectionists. In common with other critics, they characterized antivivisectionists as silly, sentimental, shrill women. Maurice Hall charged antivivisectionists with placing an unnatural value on dogs and other traditional companion animals because they could not have normal relationships with other humans. This displacement of affection followed, critics warned, from women who remained childless, or who had experienced disappointment in love. These unnatural women, furthermore, little understood the importance of scientific breakthroughs; they were impractical and irrational.[28]

The gendered slant of these arguments came easily to veterinarians, whose profession had been born of a hard-nosed utilitarianism and self-conscious masculinity. As applied scientists, veterinarians saw themselves as rationalists; in laboratories and barnyards, their profession was eminently useful and practical and situated squarely in the province of men. This vision also described the personnel of the profession; deans of veterinary schools (until the 1970s) and the American Veterinary Medical Association (until the 1940s) regularly and openly denied women admission. Veterinarians felt themselves to be humanists, but the animal rights position alienated them, and the potential social disruption implicit in antivivisectionists' goals frightened them. By marginalizing antivivisectionists personally and rhetorically, veterinarians made morality consistent with Americans' practice of assigning value to animals based on practicality.

Valuing Animals in the Twenty-first Century

The patterns of human-animal relations established a century ago remain with us today because they continue to reconcile morality with animal use. Yet it is clear that they are constantly subject to revision, just as they have been for the past hundred years. Food-producing animals, for example, continue to be most valuable as cheap sources of milk, meat, and the other raw materials that find their way into most of the processed foods that Americans consume. Keeping the price of food low, however, did not prevent periodic flareups of consumer concern about food wholesomeness in the 1970s through 1990s. Old debates about federal government or industry responsibility for food safety resurfaced and changed procedures within both the barnyard and government agencies.

In 1993, one of a chain of national fast food restaurants was the source of *Escherichia coli* bacteria in undercooked hamburgers that killed several children in the Pacific Northwest. *E. coli* infections, scientists allege, doubled in number between 1995 and 2000. The Center for Science in the Public Interest, one of several interested consumer groups, helped to push for a federal law that would be a "landmark in improving meat and poultry safety" by adding microbial tests for *E. coli* to federal meat inspectors' century-old "poke and sniff" methods. The new federal rules, implemented in 1996, made the greatest changes to American meat inspection law since

1906. As with the 1906 legislation, the new rules also represented the final product of months of negotiation between consumer groups and livestock producers (such as the National Cattlemen's Beef Association) and the meat-packing industry.[29]

Consumer groups who valued the health of animals and animal products in the last quarter of the twentieth century often based their arguments on the theme that animal producers had consistently emphasized over the preceding decades: healthy animals and animal products came from creatures living in conditions conducive to their physical and mental welfare. Any taint in the food supply, therefore, implicated animals' living conditions and accused producers of tampering with animals' "natural" needs and life cycles. The twentieth-century "improvements" in livestock husbandry that supported the value of animals in large numbers drew criticism from those concerned with the welfare of animals and human consumers alike. As a scientist from the advocacy group Environmental Defense reminded *New York Times* readers in 2001, the antibiotics that led to *E. coli* proliferation and consumers' illnesses had been fed to confinement-raised animals as "compensation for unsanitary, stressful conditions in crowded factory farms."[30]

In a similar example, veterinarians and other scientists at Monsanto, the leading agricultural biotechnology company in the United States at the end of the twentieth century, tested a laboratory-produced version of a bovine somatotropin (BST, a growth hormone). They found that injecting BST into dairy cows caused them to produce between 10 and 25 percent more milk than untreated cows. In 1994, when the first milk from cows injected with the genetically engineered hormone went on sale, critical consumer groups not only voiced distrust of the scientists who assured them of the hormone's safety, but also expressed anger that milk, "a pure food for innocents," had been targeted for "unnatural" improvement. Indeed, small dairy owners fearful of being driven out of business by much larger competitors seized the chance to advertise their milk as having been produced "naturally."[31]

The debate over the "natural" in animal production gained urgency with the British revelation in 1996 that bovine spongiform encephalopathy (BSE or "mad cow disease") had probably been transmitted to Britons who ingested infected animal products. Scientists traced BSE to the food-animal production practices of the 1970s and '80s. As a way to increase the animals' output of milk and growth rate, producers fed them high-protein diets containing the remains of animals infected with BSE or other transmissible encephalopathies. American critics of the food-production system argued that treating these animals unnaturally—"making herbivores eat the remains of their sisters"—violated nature's laws. By doing so, humans invited "nature's revenge": disease in one of its more fascinating and terrifying forms. The fact that BSE did not physically affect people living in the United States made no difference to the rhetorical power of this argument.[32] Searching for a meaning behind this pernicious zoonotic disease, critics have challenged more than just the practices of the food-production systems in the United Kingdom and the United States. They have reexamined Americans' relationships with their domesticated creatures in terms of humans' proper respect for obeying natural laws. Placing domesticated animals squarely into the realm of the "natural," such critiques underscore the potential for unimpeded human use to violate moral imperatives, with dire consequences.

If some domesticated animals function as representatives of natural law, others challenge human law. As a legal concept, animal value is in the midst of a fundamental transformation in the United States. Currently focusing on pet animals, this transformation challenges the basis of some animals' long-standing legal identity as property, or chattel. As a first step, the value of animals as emotional companions has been judicially and legislatively recognized in isolated cases for the first time in U.S. history. In 2000, for example, the state of Tennessee passed a statute that allowed courts to routinely award emotional distress damages to an animal's owner after its loss owing to "the negligent act of another." From another angle, advocates for treating animals as autonomous beings (a concept congenial to a rights philosophy) have helped to pass ordinances in three U.S. cities that redefine animal "owners" as animal "guardians." Although the rhetorical power of such a change far exceeds its legal power, the "guardian" ordinances have been widely interpreted as components of "a growing recognition of pets as more than property."[33] Companion animals have become the test case for a legal reassessment of the value of domestic animals in American society.

This should not be surprising, since domesticated animals have been invested with complex profiles of value for much of the country's history. Veterinarians over the past century have worked to shape standards of animal care, thus mediating the ways that Americans have interacted with domesticated creatures. Conversely, veterinary science itself developed within a system of animal value, with its intellectual and sociological characteristics dependent on the worth assigned to its patient populations. Such a system has incorporated many social needs and cultural beliefs within the economic and legal worth placed on animals. It has also allowed Americans to balance their use of animals with their need to see themselves as humanitarians. The failure to do so would represent a significant challenge to social order and cultural complacency in the twenty-first century, just as it would have a hundred years earlier.

Notes

The following acronyms are used in the notes: USDA, U.S. Department of Agriculture; RG, Record Group; NARA, National Archives and Records Administration.

Introduction

1. In deeming themselves experts on animal bodies, behavior, health, and disease, veterinarians often paralleled allopathic physicians, who claimed proprietary knowledge of human bodies and minds by the end of the nineteenth century.

2. On the professionalization of veterinary medicine, see J. F. Smithcors, *The American Veterinary Profession: Its Background and Development* (Ames: Iowa State University Press, 1963), and Bert W. Bierer, *A Short History of Veterinary Medicine in America* (East Lansing: Michigan State University Press, 1955). The term *companion animal* is preferred by many late-twentieth-century animal advocates over *pet*, which they perceive as trivializing the human-animal relationship and connoting oppressive ownership.

3. E. P. Thompson, "The Moral Economy of the English Crowd in the Eighteenth Century" and "The Moral Economy Revisited," in *Customs in Common* (New York: New Press, 1991); Donald Worster, *Nature's Economy: A History of Ecological Ideas*, 2nd ed. (Cambridge: Cambridge University Press, 1995).

4. Robert Wiebe, *The Search for Order, 1877–1920* (New York: Hill & Wang, 1967); William E. Leuchtenburg, *The Perils of Prosperity, 1914–1932* (Chicago: University of Chicago Press, 1958); Clay McShane, *Down the Asphalt Path: The Automobile and the American City* (New York: Columbia University Press, 1994); Joel Tarr, "Urban Pollution—Many Long Years Ago," *American Heritage* 22 (October 1971): 65–69, 106; and Martin Melosi, *The Sanitary City: Urban Infrastructure in America from Colonial Times to the Present* (Baltimore, MD: Johns Hopkins University Press, 2000).

5. James Serpell, *In the Company of Animals: A Study of Human-Animal Relationships* (Cambridge: Cambridge University Press, 1986); Jennifer Wolch and Jody Emel, eds., *Animal Geographies: Place, Politics, and Identity in the Nature-Culture Borderlands* (London: Verso, 1998); Yi-Fu Tuan, *Dominance and Affection: The Making of Pets* (New Haven, CT: Yale University Press, 1984); F. Barbara Orlans, Tom L.

Beauchamp, Rebecca Dresser et al., *The Human Use of Animals: Case Studies in Ethical Choice* (Oxford: Oxford University Press, 1998); Keith Thomas, *Man and the Natural World: A History of the Modern Sensibility* (New York: Pantheon Books, 1983); Harriet Ritvo, *The Animal Estate: The English and Other Creatures in the Victorian Age* (Cambridge, MA: Harvard University Press, 1987); Kathleen Kete, *The Beast in the Boudoir: Petkeeping in Nineteenth Century Paris* (Berkeley: University of California Press, 1994); Lawrence Finsen and Susan Finsen, *The Animal Rights Movement in America: From Compassion to Respect* (New York: Twayne Publishers, 1994); Susan E. Lederer, "The Controversy over Animal Experimentation in America, 1880–1914," in Nicolaas A. Rupke, ed., *Vivisection in Historical Perspective* (London: Routledge, 1990); Harriet Ritvo, "Plus ça Change: Antivivisection Then and Now," *Science, Technology, and Human Values* 9 (Spring 1984): 57–66.

6. Elliott West, *The Way to the West: Essays on the Central Plains* (Albuquerque: University of New Mexico Press, 1995), p. 52. For a discussion of animal agency and scholarly critiques of anthropomorphism, see Chris Philo and Chris Wilbert, "Animal Spaces, Beastly Places: An Introduction," in Philo and Wilbert, eds., *Animal Spaces, Beastly Places: New Geographies of Human-Animal Relations* (London: Routledge, 2000), pp. 14–23; Mary Midgley, *Animals and Why They Matter: A Journey Around the Species Barrier* (Harmondsworth: Penguin, 1983), chap. 11.

7. Georg Simmel, *The Philosophy of Money*, David Frisby, ed., trans. Tom Bottomore and David Frisby from a first draft by Kaethe Mengelberg, 2nd ed. (London: Routledge, 1990), quote at p. 56. Viviana Zelizer, *Pricing the Priceless Child: The Changing Social Value of Children* (Princeton: Princeton University Press, 1994) (first published by Basic Books in 1985).

8. Virginia DeJohn Anderson, *Creatures of Empire: People and Animals in Early America* (forthcoming).

9. Max Weber, "Bureaucracy," in H. H. Gerth, ed., and C. Wright Mills, trans., *From Max Weber: Essays in Sociology* (London: Routledge, 1948), pp. 196–244; Peter Wagner, *A Sociology of Modernity: Liberty and Discipline* (London: Routledge, 1994).

10. Charles E. Rosenberg, "Introduction," in Charles E. Rosenberg and Janet Golden, eds., *Framing Disease: Studies in Cultural History* (New Brunswick, NJ: Rutgers University Press, 1992), p. xxiii.

Chapter 1. Doctoring a Nation of Animals at the Century's Turn

1. Private schools especially varied; some won praise for strict matriculation requirements and longer courses; others were criticized as mere business ventures. See editorial, "McKillip Veterinary College," *Journal of Comparative Medicine and Veterinary Archives* 15 (October 1894): 353; editorial, "A Costly Experience," *Journal of Comparative Medicine and Veterinary Archives* 17 (November 1896): 788–789; editorial, "What Will the Others Do?" *Journal of Comparative Medicine and Veterinary Archives* 17 (June 1896): 481–482.

2. Department of the Interior, Census Office, *Special Census Report on the Occupations of the Population of the United States at the Eleventh Census, 1890* (Washington, DC: Government Printing Office, 1896), pp. 11–13; *2001 AVMA Membership Directory*

and *Resource Manual* (Schaumburg, IL: American Veterinary Medical Association, 2001), pp. 212–227; *1999 AVMA Membership Directory and Resource Manual* (Schaumburg, IL: American Veterinary Medical Association, 1999), pp. 187–204; J. F. Smithcors, *The American Veterinary Profession: Its Background and Development* (Ames: Iowa State University Press, 1963), pp. 81–112, 147–148. Of course, many animal owners cared for their animals themselves; see Smithcors, *American Veterinary Profession,* pp. 187–214.

3. R. R. Dykstra, "The Farmer and the Veterinarian," *Journal of the American Veterinary Medical Association* 69 (April–September 1926): 343–356; for more on stable culture and stereotypes of grooms, see "Jorrocks" [James Albert Garland], *The Private Stable* (Boston: Little, Brown, 1899).

4. R. J. Dinsmore, *"Hoss" Doctor* (Boston: Waverly House, 1940), p. 15.

5. Dinsmore, *"Hoss" Doctor,* p. 13.

6. Jennings quoted in Smithcors, *American Veterinary Profession,* p. 284.

7. U. G. Houck, *The Bureau of Animal Industry of the United States Department of Agriculture: Its Establishment, Achievements, and Current Activities* (Washington, DC: published by the author, 1924); Fred W. Powell, *The Bureau of Animal Industry: Its History, Activities, and Organizations* (Baltimore, MD: Johns Hopkins Press, 1927), pp. 34–36.

8. See deliberations on December 3, 1883, 1st Session, 48th U.S. Congress, on House of Representatives Bill 3967; see also Vivian Wiser, Larry Mark, and H. Graham Purchase, *100 Years of Animal Health, 1884–1984* (Beltsville, MD: Associates of the National Agricultural Library, 1987).

9. The culture of animal doctoring is only one of many "institutions and relations [that have] become gendered"; see Ava Baron, ed., *Work Engendered: Toward a New History of American Labor* (Ithaca, NY: Cornell University Press, 1991), p. 36.

10. The masculine culture of the livery stable continued into the twentieth century; see Paul C. Johnson, *Farm Animals in the Making of America* (Des Moines, IA: Wallace Homestead, 1975), pp. 27, 29, 33.

11. On American women physicians, see e.g. Regina Markell Morantz-Sanchez, *Sympathy and Science: Women Physicians in American Medicine* (New York: Oxford University Press, 1985).

12. Quoted in Roscoe Bell, editorial, *American Veterinary Review* 21 (1897): 595–596.

13. Charles E. Rosenberg and Carroll Smith-Rosenberg, "The Female Animal: Medical and Biological Views of Women," in Charles E. Rosenberg, *No Other Gods: On Science and American Social Thought* (Baltimore, MD: Johns Hopkins University Press, 1976), chap. 2.

14. For more information about women veterinarians, see Phyllis Hickney Larsen, ed., *Our History of Women in Veterinary Medicine: Gumption, Grace, Grit and Good Humor* (Madison, WI: Omnipress, for the Association for Women Veterinarians, 1997).

15. Quoted in Roscoe Bell, editorial, *American Veterinary Review* 21 (1897): 595–596, quote at p. 596.

16. From a 1915 editorial, quoted in J. F. Smithcors, "Women Veterinarians: Yes-

terday, Today, Tomorrow," *Modern Veterinary Practice* 55 (December 1974): 933–940, quote at p. 933.

17. Charles E. Rosenberg, "The Therapeutic Revolution: Medicine, Meaning, and Social Change in Nineteenth Century America," in Morris Vogel and Charles E. Rosenberg, eds., *The Therapeutic Revolution: Essays on the Social History of American Medicine* (Philadelphia: University of Pennsylvania Press, 1979), chap. 1; Martin Pernick, *A Calculus of Suffering: Pain, Professionalism, and Anesthesia in Nineteenth-Century America* (New York: Columbia University Press, 1985), pp. 26–34.

18. An 1895 fee schedule published by the Maine Veterinary Medical Association listed sixteen common operations; the procedures described here represent nine of them. Animal doctors sometimes became specialists in one procedure; see an interesting account of a late nineteenth-century castration expert in Smithcors, *American Veterinary Profession*, pp. 273–278, 484.

19. For examples of other disciplines using educational institutions as a professionalizing strategy, see Samuel Haber, *The Quest for Authority and Honor in the American Professions, 1750–1900* (Chicago: University of Chicago Press, 1991). On land-grant colleges and agricultural experiment stations, whose intellectual turf overlapped with that of veterinary medicine, see Charles E. Rosenberg, *No Other Gods*, chaps. 9 and 10; Alan I. Marcus, *Agricultural Science and the Quest for Legitimacy: Farmers, Agricultural Colleges, and Experiment Stations, 1870–1890* (Ames: Iowa State University Press, 1985); Jeffrey W. Moss and Cynthia B. Lass, "A History of Farmers' Institutes," *Agricultural History* 62 (1988): 150–163.

20. A. S. Copeman, *Introductory Lecture, Delivered at the Opening of the New York College of Veterinary Surgeons, November 6, 1865* (New York: M. T. Tyler, 1865), p. 5.

21. Pierre A. Fish, untitled essay on history of veterinary medicine, ca. 1925, in Ellis P. Leonard, ed., *Pierre A. Fish: A Collection of Historical Papers and Memorabilia Assembled for the Archives of the Flower and Olin Libraries, Vol. 3* (Ithaca, NY: Flower Veterinary Library, 1983), pp. 450–451.

22. For a discussion of the "monopoly of competence," see Magali Sarfatti Larson, *The Rise of Professionalism: A Sociological Analysis* (Berkeley: University of California Press, 1977), pp. 221–224.

23. Faculty of the School of Veterinary Medicine, University of Pennsylvania, *The Veterinary Profession: Its Relation to the Health and Wealth of the Nation* (Philadelphia, PA: Avil Printing, 1897).

24. University President Jacob Gould Schurman, "The State Veterinary College," excerpt from annual report, Cornell University, November 14, 1894, Flower-Sprecher Library Archives, Cornell University, Ithaca, NY.

25. Alexandre Liautard, editorial, *American Veterinary Review* 3 (1879): 21. As editor of the review for more than 20 years, Liautard's views were widely disseminated.

26. U.S. Department of Commerce, Bureau of the Census, *United States Summary, Population Bulletin* (Washington, DC: Government Printing Office, 1931), pp. 4, 7. Historian Philip M. Teigen first approached domestic animal demographics by using measurements of density (animals per square mile) in Teigen, "'She Ain't

What She Used to Be': Revisionist Views of the Horse in America, 1860–1920," conference paper presented at a meeting of the American Veterinary History Society, in conjunction with the annual meeting of the American Association for the History of Medicine, April 3, 1997; I am indebted to him for this concept.

27. Mary Neth, *Preserving the Family Farm: Women, Community, and the Foundations of Agribusiness in the Midwest, 1900–1940* (Baltimore, MD: Johns Hopkins University Press, 1995); John L. Shover, *First Majority, Last Minority: The Transforming of Rural Life in America* (DeKalb: Northern Illinois University Press, 1976); David B. Danbom, *Born in the Country: A History of Rural America* (Baltimore, MD: Johns Hopkins University Press, 1995), esp. chap. 7 and pp. 112, 194.

28. Census Office, *Special Census Report on the Occupations . . . 1890*, pp. 11–13.

29. Lizabeth Cohen, *Making a New Deal: Industrial Workers in Chicago, 1919–1939* (Cambridge: Cambridge University Press, 1991), p. 28; U.S. Department of Commerce and Labor, Census Office, *Census Reports, Volume VII, Twelfth Census, 1900: Manufactures, Part II* (Washington, DC: U.S. Census Office, 1902), p. 184; U.S. Department of Commerce and Labor, Census Office, *Census Reports, Volume VII, Twelfth Census, 1900, Manufactures, Part I* (Washington, DC: U.S. Census Office, 1902), p. cclxv; William Cronon, *Nature's Metropolis: Chicago and the Great West* (New York: W. W. Norton, 1991), p. 211.

30. Philip M. Teigen, personal communication, May 7, 1997.

31. For more challenges to popular representations of the western United States, see Patricia Nelson Limerick, *The Legacy of Conquest: The Unbroken Past of the American West* (New York: W. W. Norton, 1987). Historian Joel Tarr has been a leading proponent of the study of urban horses; see Tarr, "Urban Pollution—Many Long Years Ago," *American Heritage* 22 (October 1971), pp. 65–69, 106.

32. Goats were locally popular in a few places, including Queens, NY; New Orleans, LA; Milwaukee, WI; and Oakland and San Francisco, CA. Census Office, *Bulletin No. 17, Domestic Animals in Barns and Inclosures, Not on Farms or Ranges* (Washington, DC: Census Office, 1900), pp. 3, 96–98.

33. Charles Dickens, *American Notes* (Oxford: Oxford University Press, 1987 [1842]), pp. 86, 87; Census Office, *Domestic Animals in Barns*, pp. 1, 28; Department of Commerce, Bureau of the Census, *Fourteenth Census of the United States, Taken in the Year 1920, Volume V: Agriculture* (Washington, DC: Government Printing Office, 1922), p. 622; see tables I and II also. These population numbers are cross-sectional; seven to ten million hogs traveled through the Chicago stockyards in a single year.

34. Bureau of Municipal Research of Philadelphia, *Municipal Street Cleaning in Philadelphia* (Philadelphia: Bureau of Municipal Research, June 1924), pp. 51–52. Garbage collectors in smaller cities and towns especially benefited from using local pigs as garbage disposals. For the procedures of towns and cities in New York State and veterinary recommendations, see Raymond Russell Birch Papers, 24/2/525, folders dated 1912–1918, Rare and Manuscript Collections, Carl A. Kroch Library, Cornell University Library, Ithaca, NY.

35. Census Office, *Domestic Animals in Barns*, p. 29; Bureau of the Census, *Four-*

teenth Census, Volume V, p. 622; see tables I and II also. Again, these population numbers are cross-sectional.

36. Bureau of the Census, *Fourteenth Census, Volume V*, p. 614. The war years, 1917 to early 1919, accounted for much of the 26% increase. Census Office, *Twelfth Census of the United States, Taken in the Year 1900, Manufactures, Part III* (Washington, DC: Government Printing Office, 1902), p. 406.

37. While the midwestern slaughtering centers such as Chicago were packers, generating many products from each animal, the center of the fresh meat slaughtering business was in the cities and towns of New York State. Most of both enterprises' products were consumed by citizens of the large eastern cities. See Cronon, *Nature's Metropolis*, esp. chap. 5.

38. Caroline Hedger, "The Unhealthfulness of Packingtown," *World's Work* (May 1906): 7507–7510, quotes at p. 7510; Harold L. Platt, "Jane Addams and the Ward Boss Revisited: Class, Politics, and Public Health in Chicago, 1890–1930," *Environmental History* 5 (April 2000): 194–222.

39. Henry E. Alvord and Raymond A. Pearson, *The Milk Supply of Two Hundred Cities and Towns*, USDA, *Bureau of Animal Industry Bulletin No. 46* (Washington, DC: Government Printing Office, 1903), pp. 13–14, 26, 30, 34, 40.

40. Bureau of the Census, *Fourteenth Census, Volume V*, pp. 621, 625, 626. Cities and towns with human populations of less than 25,000 did tend to maintain their numbers of dairy cows, so that the census registered an overall increase of 4% in urban dairy cows between 1910 and 1920. C. Hampson Jones, "The Present Needs of the Milk Supply of Baltimore," *Charities* 16 (August 4, 1906): 499; Alvord and Pearson, *Milk Supply*, pp. 26, 28. Veterinarians concerned with reforming animal production methods to fit rationalized cities and suburbs correspond with Daniel T. Rodgers' category of reformers interested in implementing ideas of social efficiency; see Rodgers, "In Search of Progressivism," *Reviews in American History* 10 (December 1982): 113–132, esp. pp. 122–123.

41. William H. Wilson, *The City Beautiful Movement* (Baltimore, MD: Johns Hopkins University Press, 1989); Joel Tarr, *The Search for the Ultimate Sink: Urban Pollution in Historical Perspective* (Akron, OH: University of Akron Press, 1996); Martin Melosi, *The Sanitary City: Urban Infrastructure in America from Colonial Times to the Present* (Baltimore, MD: Johns Hopkins University Press, 2000), esp. chaps. 6–9. The "rationalization" of the city follows from the views of Max Weber, which are summarized well by Peter Wagner, "Sociological Reflections: The Technology Question during the First Crisis of Modernity," in Mikael Hård and Andrew Jamison, *The Intellectual Appropriation of Technology: Discourses on Modernity, 1900–1939* (Cambridge, MA: MIT Press, 1998), chap. 9.

42. Edwin Tennery Brewster, "A City of 4,000,000 Cats," *McClure's* 39 (May 1912): 54–64, quote at p. 55. Brewster was sympathetic to the SPCA and its humane mission, and the article was intended to be educational.

43. Rene Laidlaw, "Dogs in Town," *Independent* 74 (March 13, 1913): 571–573; H. D. Jones, "Dogs as Auxiliary Policemen," *World To-Day* 8 (May 1905): 551–552.

44. U.S. Department of Commerce, Bureau of the Census, *Thirteenth Census,*

1910, *Volume V: Agriculture* (Washington, DC: Government Printing Office, 1914), pp. 422–23; Philip M. Teigen, "Historical Zoogeography of the Urban Horse in the United States, 1860–1920," paper presented at the 25th Symposium of the International Committee for the History of Technology, Lisbon, Portugal, August 1998; Clay McShane and Joel A. Tarr, "The Centrality of the Horse in the Nineteenth-Century American City," in Raymond A Mohl, ed., *The Making of Urban America*, 2nd ed. (Wilmington, DE: Scholarly Resources, 1988), pp. 105–130.

45. Joel A. Tarr, "Urban Pollution—Many Long Years Ago," *American Heritage* 22 (October 1971): 65–69, 106; Nancy Tomes, *The Gospel of Germs: Men, Women, and the Microbe in American Life* (Cambridge, MA: Harvard University Press, 1998), pp. 105, 108–109.

46. Veterinary viewpoints on rabies varied, with many practitioners supporting sportsmen and pet owners in opposing regulation of dogs. James Law's 1900 report was, however, the profession's official stance on the subject; see J. F. Smithcors, "The History of Some Current Problems in Animal Disease—VII. Rabies," *Veterinary Medicine* 53 (1958): 149, 267, 435.

47. The details of hog cholera control and veterinarians' role at this time are discussed in W. B. Niles, "Field Tests with Serum for the Prevention of Hog Cholera," pp. 177–217 and A. D. Melvin, "The Control of Hog Cholera by Serum Immunization," pp. 219–224 in U.S. Department of Agriculture (hereafter USDA), *Twenty-fifth Annual Report of the Bureau of Animal Industry, 1908* (Washington, DC: Government Printing Office, 1910); Smithcors, *American Veterinary Profession*, pp. 451–457.

48. James A. Toman and Blaine S. Hays, *Horse Trails to Regional Rails: The Story of Public Transit in Greater Cleveland* (Kent, OH: Kent State University Press, 1996), pp. 21–22; Christine Meisner Rosen, *The Limits of Power: Great Fires and the Process of City Growth in America* (New York: Cambridge University Press, 1986), pp. 176–179.

49. Wilson, *The City Beautiful Movement;* Suellen Hoy, *Chasing Dirt: The American Pursuit of Cleanliness* (New York: Oxford University Press, 1995); Martin Melosi, *Garbage in the Cities: Refuse, Reform, and the Environment, 1880–1980* (College Station: Texas A&M University Press, 1981).

50. Pierre A. Fish, "Seventy Years of Veterinary Service in the United States," *Journal of the American Veterinary Medical Association* 27 (1929): 915–917; Census Office, *Census Reports, Volume II, Twelfth Census, Population, Part II* (Washington, DC: Government Printing Office, 1905), pp. 510–541; Bureau of the Census, *Thirteenth Census, Volume IV, Population, 1910: Occupation Statistics* (Washington, DC: Government Printing Office, 1914), see tables II and III on pp. 96–331; Philip M. Teigen, "Nineteenth-Century Veterinary Medicine as an Urban Profession," *Veterinary Heritage* 23 (May 2000): 1–5.

51. USDA, *Yearbook of the Department of Agriculture, 1907* (Washington, DC: Government Printing Office, 1908), p. 710; USDA, *Yearbook of the Department of Agriculture, 1898* (Washington, DC: Government Printing Office, 1899), pp. 702–703; John S. Billings, *Report on the Social Statistics of Cities in the United States at the Eleventh Census: 1890* (Washington, DC: Government Printing Office, 1895), p. 44.

52. Dykstra, "The Farmer and the Veterinarian," pp. 343–356; A. F. Schalk,

"Undeveloped Resources," *Journal of the American Veterinary Medical Association* 59 (July 1921): 410–433; L. H. Pammel, "The Education of the Veterinarian and His Relation to the Commonwealth," *American Veterinary Review* 30 (1906): 203–219 (see p. 209 for curricula).

53. Philip M. Teigen and Sheryl A. Blair, "The Massachusetts Veterinary Profession, 1882–1904," *Historical Journal of Massachusetts* 25 (1997): 63–73.

54. Arthur D. Goldhaft, *The Golden Egg* (New York: Horizon Press, 1957), p. 127.

55. A. H. Streeter, "Veterinary Practice in the East," in Faculty of the School of Veterinary Medicine, University of Pennsylvania, *The Veterinary Profession*, p. 41; A. B., V.S., "The Future of the Horse," letter to the editor, *American Veterinary Review* 23 (April 1899–March 1900): 318.

56. Goldhaft, *Golden Egg*, pp. 126–135, quote at p. 126.

57. William Herbert Lowe, "The Relation of Veterinary Medicine to the Public Health," *American Veterinary Review* 24 (1901): 588–592, quote at p. 589.

58. The intellectual and institutional loyalties of these veterinarians require more study to illuminate how germ ideas and experimental practices were transferred to the American veterinary profession. See Michael Worboys, *Spreading Germs: Disease Theories and Medical Practice in Britain, 1865–1900* (Cambridge: Cambridge University Press, 2000); John Harley Warner, *Against the Spirit of the System: The French Impulse in Nineteenth-Century American Medicine* (Princeton, NJ: Princeton University Press, 1998); and John Andrew Mendelsohn, "Cultures of Bacteriology: Foundation and Transformation of a Science in France and Germany, 1870–1914," Ph.D. diss., Princeton University, Princeton, NJ, 1996.

59. See the first two volumes of the *American Veterinary Review* particularly for Liautard's editorial views (1877). For more on Liautard and his journal, see Smithcors, *American Veterinary Profession*, chaps. 10 and 11.

60. For more on Law's role at Cornell, see Ellis Pierson Leonard, *A Cornell Heritage: Veterinary Medicine, 1868–1908* (Ithaca: Vail-Ballou Press for the New York State College of Veterinary Medicine, 1979) and *In the James Law Tradition, 1908–1948* (Ithaca: Vail-Ballou Press for the New York State College of Veterinary Medicine, 1982). Law had also studied in Joseph Lister's laboratory prior to his emigration to the United States.

61. For models of veterinary education and Cornell as an important exemplar, see Philip M. Teigen, "Getting Beyond the Stable Door: Veterinary Education in Canada and the United States, 1866–1930," paper presented at the C. A. V. Barker Symposium on Canadian Veterinary History, Ontario Veterinary College, University of Guelph, June 15, 2001.

62. Wiser et al., *100 Years of Animal Health*, pp. 169–170; Rudolf A. Clemen, *The American Meat and Livestock Industry* (New York: Ronald Press, 1923), chap. 15; Edward W. Perry, *First Annual Report, United States Bureau of Animal Industry* (Washington, DC: U.S. Government Printing Office, 1884); "Report of the Secretary: Bureau of Animal Industry," in USDA, *Yearbook of the Department of Agriculture, 1897* (Washington, DC: Government Printing Office, 1898), p. 247. The best single source for information about animal epizootics is Bert W. Bierer, *History of Animal*

Plagues of North America (printed by the author, © 1939, reprinted in Washington, DC: U.S. Department of Agriculture, 1974, pp. 17–31).

63. The history of the Bureau of Animal Industry (BAI) is detailed in Houck, *The Bureau of Animal Industry;* Fred W. Powell, *The Bureau of Animal Industry: Its History, Activities, and Organizations* (Baltimore, MD: Johns Hopkins Press, 1927); and Wiser et al., *100 Years of Animal Health.* For a summary of the specific events leading to the founding of the BAI, see Wiser et al., *100 Years of Animal Health,* p. 170. Not all livestock owners were enthusiastic about the BAI; one of the BAI's proposed activities, federal meat inspection, promised to upset the usual procedures of meat packers and the middlemen who provided them with animals. See Suellen Hoy and Walter Nugent, "Public Health or Protectionism? The German-American Pork War, 1880–1891," *Bulletin of the History of Medicine* 63 (1989): 198–224.

64. Edward W. Perry, *First Annual Report, United States Bureau of Animal Industry* (Washington, DC: Government Printing Office, 1884); Wiser et al., *100 Years of Animal Health,* p. 3.

65. Cattlemen had long suspected the ticks, but scientists had discounted the theory; see Susan D. Jones, "Laboratory Science and Common Sense," *Veterinary Heritage* 22 (November 1999): 25–30.

66. Theobald Smith and F. L. Kilborne, "Investigations into the Nature, Causation, and Prevention of Texas or Southern Cattle Fever," *USDA Bureau of Animal Industry Bulletin No. 1* (Washington, DC: Government Printing Office, 1893).

67. Salmon was most interested in developing a vaccine; veterinarian Mark Francis in Texas was the strongest proponent of cattle-dipping programs.

68. On the development of germ theories, see Worboys, *Spreading Germs; Journal of the History of Medicine and Allied Sciences* 52 (January 1997), especially Nancy J. Tomes and John Harley Warner, "Introduction to Special Issue on Rethinking the Reception of the Germ Theory of Disease: Comparative Perspectives," pp. 7–16, and Terrie M. Romano, "The Cattle Plague of 1865 and the Reception of 'The Germ Theory' in Mid-Victorian Britain," pp. 51–80.

69. Veterinary practitioners' concerns about the laboratory sciences paralleled those of physicians; see John Harley Warner, "The Fall and Rise of Professional Mystery: Epistemology, Authority, and the Emergence of Laboratory Medicine in Nineteenth-Century American," in Andrew Cunningham and Perry Williams, eds., *The Laboratory Revolution in Medicine* (Cambridge: Cambridge University Press, 1992), pp. 110–141. Anticontagionism was a belief maintained by many apprenticeship-trained veterinarians as well as by those who had been trained at the Royal Veterinary College in London throughout the nineteenth century. Philip M. Teigen has also identified particular veterinary schools whose leaders preferred to teach veterinary practice as an art based on experience, rather than as a product of applied research; see Teigen, "Getting Beyond the Stable Door."

70. F. S. Billings, "The Untrustworthiness of the Reports of the Government in Relation to Animal Diseases," *Journal of Comparative Medicine and Veterinary Archives* 12 (1891): 415–419; Smith and Kilborne, "Investigations into the Nature, Causation, and Prevention of Southern Cattle Fever," pp. 177–304. J. F. Smithcors has proposed

that Billings's challenges "insured the exactitude of the work conducted by the BAI scientists," in *American Veterinary Profession*, p. 455, and in "The Development of Veterinary Medical Science: Some Historical Aspects and Prospects," in *Advances in Veterinary Science* 9 (1964): 1–32. Bert W. Bierer made a similar point in *A Short History of Veterinary Medicine in America* (East Lansing: Michigan State University Press, 1955), pp. 71–72.

71. As Charles Rosenberg has put it, "a lack of precise meaning has rarely interfered with the efficacy of appeals to science and the promise of its application." See Rosenberg, *The Care of Strangers: The Rise of America's Hospital System* (New York: Basic Books, 1987), pp. 9, 150, 333.

72. Houck, *The Bureau of Animal Industry*, p. 289.

73. Leonard Pearson, "Report of the Committee on Intelligence and Education," *American Veterinary Review* 34 (1908–09): 91–108, quote at p. 102; Houck, *Bureau of Animal Industry*, p. 377.

74. Goldhaft, *Golden Egg*, pp. 133–135.

Chapter 2. Valuable Patients

1. Paul Pinkerton Foster, "Helping the Work Horses," *Outing Magazine* 53 (November 1908): 168–179; Foster, "Work Horse Parades," *World Today* 10 (1907): 535–538.

2. Henry C. Merwin, "Work-Horse Parades: Information Concerning Their Objects and Management and How to Make Them Successful," pamphlet issued by the American Humane Association ca. 1908, quoted in Roswell McCrea, *The Humane Movement: A Descriptive Survey* (New York: Columbia University Press, 1910), pp. 109–112; "The Philadelphia Work Horse Parade," *American Veterinary Review* 32 (November 1907): 290; P. P. Foster, "Work Horse Parades," *World To-Day* 10 (May 1906): 535–538; Alfred Stoddart, "Work Horse Parades," *Collier's* 47 (June 10, 1911): 17–18.

3. Recent social histories of the United States between 1900 and 1920 have exposed contradictions in reform efforts; reformers often ended up supporting or using the very institutions they sought to tear down. Most animal protectionists also fit this model; they promoted the humane treatment of animals sacrificed to work or slaughtered for food without questioning the necessity for these uses or the volume of human consumption of animal resources. See Susan Strasser, *Satisfaction Guaranteed: The Making of the American Mass Market* (New York: Pantheon Books, 1989); Daniel Horowitz, *The Morality of Spending: Attitudes Toward the Consumer Society in America, 1875–1940* (Baltimore, MD: Johns Hopkins University Press, 1985); Daniel T. Rodgers, "In Search of Progressivism," *Reviews in American History* 10 (December 1982): 113–132.

4. Foster, in "Helping the Work Horses," asserted that the parades "have shown that people really do care," and that they encouraged "drivers who are fond of their horses and proud of their good appearance" (p. 179).

5. McCrea, *Humane Movement*, table I; Foster, in "Helping the Work Horses," also described veterinarians volunteering services to humane institutions.

6. "Work-Horse Parade," *Rider and Driver* 62 (May 6, 1922): 15; "The Philadelphia Work Horse Parade," *American Veterinary Review* 32 (November 1907): 290.

7. Ann Greene, "More Than Iron Horses: Horses, Power, and Technology in the Nineteenth Century," Ph.D. diss. (in progress) University of Pennsylvania, Philadelphia, PA, 2002; Bureau of the Census, *Thirteenth Census of the United States, Volume IV, Population, 1910: Occupation Statistics* (Washington, DC: Government Printing Office, 1914), pp. 91–93.

8. A. B., V.S., "The Future of the Horse," letter to the editor, *American Veterinary Review* 23 (April 1899–March 1900): 318.

9. H. C. Merwin, *Road, Track, and Stable: Chapters About Horses and Their Treatment* (Boston: Little, Brown, 1892), pp. 227–228. This book, which originally appeared in installments in *Atlantic Monthly*, is organized by chapters according to horse occupations; O'Malley Knott, *Gone Away with O'Malley: Seventy Years with Horses, Hounds, and People* (New York: Doubleday, Doran, 1944), pp. 251–252. The classic literary statement on horse occupations is Anna Sewell, *Black Beauty, His Grooms and Companions: The Autobiography of a Horse* (London: Jarrold and Sons, 1877).

10. See, for example, the carriage parade described by Roy Rosenzweig and Elizabeth Blackmar, *The Park and the People: A History of Central Park* (Ithaca, NY: Cornell University Press, 1992), pp. 212–225.

11. *New York Herald*, May 5, 1860, quoted in Rosenzweig and Blackmar, *The Park and the People*, p. 216.

12. A. C. Gaylor, "The Morgan Horse," *Country Life in America* 16 (August 1909): pp. 419–420; Henry B. Rathbone, "The Story of the Morgan Horse," *Country Life in America* 17 (November 1909): 33–36. Famous horses that became national symbols include Justin Morgan, Ethan Allen, and General Knox. "The Physician's Automobile: Its Advantages and Disadvantages," *Journal of the American Medical Association* 50 (1908): 811–834; "Automobiles for Physicians' Use," *Journal of the American Medical Association* 46 (1906): 1172–1207.

13. Merwin, *Road, Track, and Stable*, pp. 229–232.

14. The best contemporary description of the lives of urban working horses is W. G. Gordon, *Horse World of London* (London: J. A. Allen, 1971; Religious Tract Society, © 1893).

15. Merwin, *Road, Track, and Stable*, pp. 227, 228, 254. For the livery cost figure, see "Jorrocks" [James Albert Garland], *The Private Stable* (Boston: Little, Brown, 1899), p. 24.

16. Foster, "Helping the Work Horses," p. 176.

17. Peter Wagner, "Sociological Reflections: The Technology Question during the First Crisis of Modernity," in Mikael Hård and Andrew Jamison, *The Intellectual Appropriation of Technology: Discourses on Modernity, 1900–1939* (Cambridge, MA: MIT Press, 1998), chap. 9. The working horse as well as motorized vehicles were understood as technologies in the nineteenth and early twentieth centuries; for more on the horse as technology, see Greene, "More Than Iron Horses."

18. Ronald R. Kline, *Consumers in the Country: Technology and Social Change in Rural America* (Baltimore, MD: Johns Hopkins University Press, 2000), p. 358.

19. "The Simonds Steam Wagon," *Scientific American* 71 (June 23, 1894): 389; "Gas Engine Tricycle," *Scientific American* 73 (January 12, 1895): 25.

20. Carl W. Gay, *Productive Horse Husbandry* (Philadelphia, PA: J. B. Lippincott, 1914), pp. 314, 318.

21. Henry Norman, "The Coming of the Automobile," *World's Work* 5 (April 1903): 3306–3307. This view equating the automobile with modern progress has been the one put forward by most historians in the voluminous literature on the automobile. For examples, see Clay McShane, *Down the Asphalt Path: The Automobile and the American City* (New York: Columbia University Press, 1994) and James J. Flink, *The Automobile Age* (Cambridge, MA: MIT Press, 1988).

22. F. H. Moore, "Horse or Automobile?" *Country Life* 17 (March 1910): 564.

23. Joseph K. Hart, "The Automobile in the Middle Ages," *Survey* 54 (August 1, 1925): 102.

24. Editorial, Cedar Rapids (Iowa) *Evening Gazette,* quoted in *Journal of the American Veterinary Medical Association* n.s. 18 (1924): 220.

25. Rene Laidlaw, "Dogs in Town," *Independent* 74 (March 13, 1913): 571. Laidlaw referred here to both horses and dogs.

26. Laidlaw, "Dogs in Town," p. 571; Leo Marx, *The Machine in the Garden: Technology and the Pastoral Ideal in America* (Oxford: Oxford University Press, 1967); Wagner, "Sociological Reflections," pp. 238–243.

27. "Growth of the Commercial Car Industry," *Literary Digest* 44 (March 9, 1912): 489; "Motor Fire-Engines," *Literary Digest* 44 (March 9, 1912): 499; for quote, see editorial, "Horse Excels Motor in Short-Haul Work," *Rider and Driver* 62 (February 25, 1922): 10–14, quote at p. 11; "The Return of the Horse," *Journal of the American Veterinary Medical Association* n.s. 65 (1924): 220.

28. A. P. Brodell and R. S. Pike, "Farm Tractors: Type, Size, Age, and Life," *U.S. Department of Agriculture, F.M. 30* (February 1942): 2; H. R. Tolley and L. M. Church, "Corn-Belt Farmers' Experience with Motor Trucks," *U.S. Department of Agriculture Bulletin No. 931* (Washington, DC: Government Printing Office, February 25, 1921); L. A. Reynoldson, W. R. Humphries, S. R. Speelman, E. W. McComas, and W. H. Youngman, "Utilization and Cost of Power on Corn-Belt Farms," *United States Department of Agriculture Technical Bulletin No. 384* (Washington, DC: Government Printing Office, October, 1933): 4–6; William L. Cavert, "Source of Power on 538 Minnesota Farms," Ph.D. diss., Cornell University, Ithaca, NY, 1929.

29. This point becomes clearer when census data on farm automobile ownership are compared with those for tractors and motor trucks. In 1930, 58% of all farms in the United States reported automobile ownership, but only 13.5% owned a tractor and 13.4% a motor truck (the two machines most likely to replace a draft horse). See Department of Commerce, Bureau of the Census, *Fifteenth Census, 1930: Agriculture, General Statistics Summary for the United States, 1929 and 1930* (Washington, DC: Government Printing Office, 1932), pp. 54–55.

30. See quote in Marvin McKinley, *Wheels of Farm Progress* (St. Joseph, MI: American Society of Agricultural Engineers, 1980), p. 79. "Have you placed a sentimental value . . . ," advertisement in *Country Gentleman* 85 (March 27, 1920): 71; "Trucks

Aren't Horses—That's For Sure!" advertisement for Dodge trucks, *Poultry Tribune* 53 (September 1947): 21; Department of Commerce, Bureau of the Census, *Fifteenth Census of the United States: The Farm Horse* (Washington, DC: Government Printing Office, 1933), p. 39.

31. Wagner, "Sociological Reflections," p. 251.

32. "Horse Excels Motor in Short-Haul Work," p. 10.

33. "The Emancipation of the Animals," *Independent* 77 (February 23, 1914): 255–256. The parallels between human and horse slave labor implied here had existed in American discourse at least since the early nineteenth century; see Greene, "More Than Iron Horses," chap. 5.

34. "Horses of Fashion and Their Homes of Luxury Amaze Uninformed Visitors and City Officials and May Result in Benefits to Their Humble Brother, the Work Horse," *Rider and Driver* 62 (May 6, 1922): 7.

35. Merwin, *Road, Track, and Stable*, pp. 227, 228, 254; "The Emancipation of the Animals," p. 256.

36. Actual numbers of horses living in nonfarm and nonrange areas must be estimated for 1930, since the census no longer counted them. *The Farm Horse* estimated the population of "city" horses to be about 750,000 in 1933 (see Department of Commerce, *The Farm Horse*, pp. 39, 52–53). The authors based this estimate on a sample of raw census figures from selected cities. The Horse Association of America estimated that there were 16,000,000 horses on farms and 1,300,000 doing nonagricultural work in 1930; see "Report of the AVMA Representative on the Advisory Board of the Horse Association of America," *Journal of the American Veterinary Medical Association* 79 (July–December 1931): 501.

37. I am indebted to Philip M. Teigen and Joel Tarr for references to the nineteenth-century replacement of horse power in manufacturing and other areas: Louis C. Hunter and Lynwood Bryant, *A History of Industrial Power in the United States, 1780–1930*, vol. 3: *The Transmission of Power* (Cambridge, MA: MIT Press, 1991); and Series S 1–14, "Total Horsepower of All Prime Movers, 1849 to 1970," from Bureau of the Census, *Historical Statistics of the United States, Part 2* (Washington, DC: Government Printing Office, 1975). For street railway data, see Department of Commerce and Labor, Bureau of the Census, *Street and Electric Railways, 1907* (Washington, DC: Government Printing Office, 1910), pp. 29, 157, 205; Alexander Easton, *A Practical Treatise on Street or Horse-Power Railways* (Philadelphia: Little, 1859).

38. See Department of Commerce, Bureau of the Census, *Fourteenth Census, 1920, Volume V: Agriculture* (Washington, DC: Government Printing Office, 1922), p. 544; USDA, *Yearbook of the Department of Agriculture, 1910* (Washington, DC: Government Printing Office, 1911), p. 628; USDA, *Yearbook of the Department of Agriculture, 1912* (Washington, DC: Government Printing Office, 1913), p. 679; USDA, *Yearbook of the Department of Agriculture, 1907* (Washington, DC: Government Printing Office, 1908), p. 710. These data are averages that included all types of horses, regardless of age or other characteristics, in the data set.

39. S. J. Schilling, "Some of the Newer Problems in Veterinary Science," *Journal*

of the *American Veterinary Medical Association* 69 (n.s. 18, April–September 1926): 569–578, esp. pp. 570–571. On the agricultural depression, see James Shideler, *Farm Crisis, 1919–1923* (Berkeley: University of California Press, 1957).

40. The availability of cheap horse meat may have helped to stimulate the expansion of the pet food industry after World War I.

41. Department of Commerce, *Farm Horse*, p. 18. This estimate may have been too low, based on the number of horsehides exported. See "Imports and Exports of Agricultural Products" in the USDA *Yearbooks of Agriculture*, 1910–1927. "The Horse as an Economic Anachronism," *Literary Digest* 47 (July 26, 1913): 142. On the meat shortage, see Report of the Secretary, "The Meat Supply," in USDA, *Yearbook of the Department of Agriculture, 1914* (Washington, DC: Government Printing Office, 1915), pp. 15–23. For a summary of the total American horse population, 1890 to 1930, see Susan D. Jones, "Animal Value, Veterinary Medicine, and the Domestic Animal Economy in the United States, 1890–1930," Ph.D. diss., University of Pennsylvania, Philadelphia, PA, 1997, table 1.1 (p. 18) and appendix A.

42. In 1900, the average adult American horse and mule were worth $53 and $65, respectively, compared with $30 for a dairy cow, $35 for a mature steer, or $3 for an adult sheep or pig. Department of the Interior, Census Office, *Census Reports, Volume V, Twelfth Census of the United States, 1900: Agriculture Part I* (Washington, DC: Government Printing Office, 1902), p. cxliii.

43. A. H. Streeter, "Veterinary Practice in the East," in University of Pennsylvania, Faculty of the School of Veterinary Medicine, *The Veterinary Profession: Its Relation to the Health and Wealth of the Nation and What It Offers as a Career* (Philadelphia: Avil Printing, 1897), p. 41; A. B., V.S., "The Future of the Horse," letter to the editor, *American Veterinary Review* 23 (April 1899–March 1900): 318.

44. A. F. Schalk, "Undeveloped Resources in Veterinary Sanitation and Hygiene," *Journal of the American Veterinary Medical Association* 59 (July 1921): 410–433, see pp. 417–418. Schalk taught the veterinary sciences at North Dakota Agricultural College. While veterinary programs such as Schalk's offered courses, they did not lead to a degree in veterinary medicine nor did they qualify students for professional practice; thus they are not the focus of the discussion here. For more on the horse-centered veterinary school curricula, see David White (dean of the Ohio State College of Veterinary Medicine), as quoted in [Richard Compton], *A Legacy for Tomorrow, 1885–1985* (Columbus: Ohio State University College of Veterinary Medicine and Biomedical Sciences, 1984), p. 36. Philip M. Teigen and Sheryl A. Blair, "The Massachusetts Veterinary Profession, 1882–1904," *Historical Journal of Massachusetts* 25 (1997): 63–73; Philip M. Teigen, "'She Ain't What She Used To Be': Revisionist Views of the Horse in America, 1860–1920," conference paper presented at the American Veterinary History Society session, in conjunction with the annual meeting of the American Association for the History of Medicine, April 3, 1997; Philip M. Teigen, "Nineteenth-Century Veterinary Medicine as an Urban Profession," *Veterinary Heritage* 23 (May 2000): 1–5. James Herriot [Alfred Wight], *All Things Bright and Beautiful* (London: BCA, 1995, © 1973), pp. 160–161.

45. "The Veterinarian's Future," quoted in editorial, *American Veterinary Review* 20 (April 1896–March 1897): 93.

46. Pierre A. Fish, "The Veterinary Outlook from a Teacher's Viewpoint," *Journal of the American Veterinary Medical Association* 61 (July 1922): 1–10, see p. 6; R. R. Dykstra, "The Farmer and the Veterinarian," *Journal of the American Veterinary Medical Association* 69 (April–September 1926): 343–356, see p. 343.

47. V. A. Moore, "Veterinary Science as an Economic Factor," *Cornell Veterinarian* 16 (1926): 31–37, quote at p. 31.

48. Dykstra, "The Farmer and the Veterinarian," pp. 343–344.

49. Historians have cited many factors in the closing of the private schools, including the horse crisis, the deaths of schools' founders, and the regulations imposed on schools by the Bureau of Animal Industry (BAI) and the War Department. See "Former Veterinary Medical Institutions in the United States," *American Veterinary Medical Association Directory* (Schaumburg, IL: American Veterinary Medical Association, 1997), p. 224; "Another College Closes," editorial, *Journal of the American Veterinary Medical Association* n.s. 18 (July 1924): 401–402.

50. R. M. Staley, "After High School—What? The Possibilities of the Veterinary Profession as a Life Work," *Journal of the American Veterinary Medical Association* n.s. 18 (1924): 49; "The Graduate Veterinarians of 1921," editorial, *Journal of the American Veterinary Medical Association* 59 (September 1921): 679–680; untitled news item, *Journal of the American Veterinary Medical Association* 69 (April–September 1926): 795; Pierre A. Fish, "A Brief Veterinary Survey," *Cornell Veterinarian* 20 (April 1930): 101–105; J. R. Mohler to C. H. Stange, February 21, 1922, Record Group (hereafter RG) 17 (Bureau of Animal Industry, Central Correspondence, 1913–1953), box 30, file 1.092, National Archives and Records Administration, College Park, MD (hereafter NARA).

51. Staley, "After High School . . . ," p. 50. Total enrollment at most veterinary colleges was lowest between 1920 and 1928, and schools associated with state universities came under pressure to attract a sufficient number of students or be eliminated. See, for example, the case of the Ohio State University College of Veterinary Medicine, in [Richard Compton], *Legacy for Tomorrow, 1885–1985*, pp. 33–36.

52. A. B, V.S., "The Future of the Horse," letter to the editor, *American Veterinary Review* (April 1899–March 1900) 23: 318.

53. "Veterinarians Becoming M.D.s," *American Veterinary Review* 20 (April 1896–March 1897): 817–818.

54. R. R. Dykstra, "The Farmer and the Veterinarian," *Journal of the American Veterinary Medical Association* 69 (April–September 1926): 343.

55. W. L. Williams, Department of Surgery and Obstetrics annual report submitted to Dean V. A. Moore, April 12, 1913, file labeled "Williams-Pound Correspondence," third-floor archive cage, Roswell P. Flower Library, New York State College of Veterinary Medicine, Cornell University, Ithaca, NY; quote appears in Denny H. Udall, "1910 to 1935," n.d., from box "D. H. Udall, Miscellaneous Speeches and Correspondence," third-floor archive cage, Roswell P. Flower Library, New York State College of Veterinary Medicine, Cornell University, Ithaca, NY.

56. Quoted by Veranus A. Moore, "The Significance of Crises in the Veterinary Profession," *Cornell Veterinarian* 15 (1925): 391.

57. "The Horseless Age," *American Veterinary Review* 20 (April 1896–March 1897): 92–93; A. B., "The Future of the Horse," p. 319.

58. See, for example, the following *American Veterinary Review* articles from the first decade of the twentieth century: "The Automobile Discredited by Its Friends," 28 (April 1904): 10; "A Refreshing Evidence of Sanity," 28 (March 1905): 1120–1123; "A Scientific Test of the Endurance of the Horse," 28 (March 1905): 1115–1118; "The Carriage Horse Coming Back," 32 (October 1907): 15.

59. Editorial, "The Future of the Horse," *American Veterinary Review* 23 (April 1899–March 1900): 319.

60. Editorial, "Recent Data in Veterinary Science—the Horse versus the Auto for the Physician," *American Veterinary Review* 32 (March 1908): 740–742; quotes in editorial, "The Carriage Horse Coming Back," [summary of a *New York Herald* article], *American Veterinary Review* 32 (October 1907): 15.

61. Editorial, "Talk Horse," *Veterinary Medicine* 16 (June 1921): quote at p. 94; editorial, "Grain Surplus and Horse Power," *Journal of the American Veterinary Medical Association* 77 (July–December 1930): 688–690.

62. Reply to A. B., "The Future of the Horse," p. 318.

63. John Harley Warner, "The Fall and Rise of Professional Mystery: Epistemology, Authority and the Emergence of Laboratory Medicine in Nineteenth-Century America," in Andrew Cunningham and Perry Williams, eds., *The Laboratory Revolution in Medicine* (Cambridge: Cambridge University Press, 1992), pp. 110–141. Although veterinary education and its defined intellectual base were well established in European schools by 1800, most trained veterinarians who organized the permanent American schools (starting in 1879) seem to have been friendly toward germ ideas. This is not to deny the existence of competing contagionist and anticontagionist theories within the body of veterinary practitioners, however, and that debate deserves more scrutiny.

64. Veterinary knowledge and practice (and public attitudes toward the profession) changed much more slowly than the above discussion might lead the reader to believe. The adoption of laboratory sciences as a basis for veterinary curricula, a slow process in itself, almost certainly did not overshadow traditional craft practices and beliefs. Rather, the "new" veterinary profession of the 1890s incorporated both.

65. L. H. Pammel, "The Education of the Veterinarian and His Relation to the Commonwealth," *American Veterinary Review* 30 (1906): 203–219. One of the earliest advocates of bacteriology in the veterinary curriculum was H. J. Detmers, who brought it in the 1870s to the curricula of the Iowa State and Ohio State schools (although he was primarily a pathologist). See C. H. Stange, "History of Veterinary Medicine at Iowa State College," pamphlet printed on the semicentennial of the school (Ames, IA: Division of Veterinary Medicine, June 1929), p. 41 and [Compton] *Legacy for Tomorrow*, pp. 7–8.

66. Cited in [Compton], *Legacy for Tomorrow*, p. 4.

67. Pammel, "The Education of the Veterinarian," p. 204.

68. F. W. Beckman, "Veterinary Education Comes into Its Own in the West," *American Veterinary Review* 42 (1912–1913): 88–92, quote at pp. 89–90. For more on the place of physiology in the veterinary curriculum, see Pierre A. Fish, "Physiology as a Fundamental in Veterinary Education," *Science* 34 (n.s.) (November 24, 1911): 700–704.

69. Schalk, "Undeveloped Resources," pp. 417–418, 427–428; W. K. Lewis, "Current Education of the Practitioner," *Journal of the American Veterinary Medical Association* 65 (1924): 39–42; Walter L. Williams, "The Problems and the Opportunities of the Veterinarian," *The Report of the Conference at the New York State Veterinary College during the Semi-Centennial Celebration of Cornell University, July 15, 1919* (Ithaca, NY: Cornell University, 1919), pp. 30–34, see p. 30.

70. Teigen, "Beyond the Stable Door," pp. 3, 4, 17.

71. Jones, "Animal Value, Veterinary Medicine, and the Domestic Animal Economy," pp. 250–255.

72. For example, BAI chief Daniel Salmon had been instrumental in the establishment of the Association of Experiment Station Veterinarians (AESV) in 1896. Salmon was also an educator (National Veterinary College, Washington, DC) as was Leonard Pearson, another founder of the AESV, who in 1900 served simultaneously as Pennsylvania's state veterinarian, an officer of the American Veterinary Medical Association, and dean of the University of Pennsylvania School of Veterinary Medicine. See "History of the Association," in *Proceedings of the Second Annual Meeting of the Association of Experiment Station Veterinarians, Bureau of Animal Industry Bulletin No. 22* (Washington, DC: Government Printing Office, 1898), p. 26.

73. "Student Enrollment for 1926–27," *Journal of the American Veterinary Medical Association* 70 (November 1926): 143–144; U. G. Houck, *The Bureau of Animal Industry of the United States Department of Agriculture: Its Establishment, Achievements, and Current Activities* (Washington, DC: published by the author, 1924), p. 371; Richard P. Lyman et al., "Report and Recommendations Regarding Veterinary Colleges in the United States," *USDA, Bureau of Animal Industry Circular 133* (Washington, DC: Government Printing Office, July 6, 1908), p. 1; David S. White, "Our Profession," *Journal of the American Veterinary Medical Association* 59 (April 1921): 5–17, see p. 9; and A. F. Schalk, "Undeveloped Resources in Veterinary Sanitation and Hygiene," *Journal of the American Veterinary Medical Association* 59 (July 1921): 410–433, see p. 421. Veterinary applicants' letters to the BAI are located in its files on education, RG 17 (Bureau of Animal Industry, Central Correspondence, 1913–1953), boxes 30 and 31, file 1.092, NARA.

74. Historian Brian Balogh has identified the USDA and its components (including the BAI) as one of the stronger federal agencies, especially after World War I. See Balogh, "Reorganizing the Organizational Synthesis: Federal-Professional Relations in Modern America," *Studies in American Political Development* 5 (Spring 1991): 119–172, esp. pp. 150–151.

75. Richard P. Lyman et al., "Report and Recommendations Regarding Veteri-

nary Colleges in the United States," *USDA, Bureau of Animal Industry Circular 133* (Washington, DC: Government Printing Office, July 6, 1908).

76. Control over the veterinary labor supply, seemingly a logistical problem, was actually a political one for the BAI; public criticism of meat inspection procedures that led to the Meat Inspection Act of 1906 had been partially focused on the BAI (see Chapter Three in this volume).

77. Lyman et al., "Report and Recommendations Regarding Veterinary Colleges in the United States"; the list of schools' ratings is on p. 10. Leonard Pearson, "Report of the Committee on Intelligence and Education," *American Veterinary Review* 34 (1908–09): 91–108, see p. 92 for the membership of the committee.

78. Charles D. Folse (Kansas City Veterinary College) to J. R. Mohler (Chief, BAI), June 27, 1917, RG 17, box 31, file 1.092, NARA.

79. Pearson, "Report of the Committee on Intelligence and Education," is the most comprehensive critique of the BAI regulations. After the Wilson committee submitted its report in July, the AVMA quickly ratified its recommendations in September. Secretary Wilson convened representatives of many veterinary colleges, including those deemed "unacceptable," for a two-day conference in January 1909, to air their concerns. Following "a full discussion of the problems" foreseen with the new regulations, the college representatives agreed to ratify them. See Houck, *Bureau of Animal Industry,* 375–376. For a more detailed discussion of the regulations applied to veterinary schools by the BAI, see Jones, "Animal Value, Veterinary Medicine, and the Domestic Animal Economy," pp. 231–236.

80. Student lists may be found in the Bureau of Animal Industry archives, RG 17, box 30, file 1.092, NARA.

81. "Regulations Governing Entrance," pp. 2–6, 8.

82. Jones, "Animal Value, Veterinary Medicine, and the Domestic Animal Economy," chap. 4.

83. The final regulations were published in "Regulations Governing Entrance to the Veterinary Inspector Examination," *USDA, Bureau of Animal Industry Circular 150,* August 9, 1909. Pearson, "Report of the Committee on Intelligence and Education," contains educators' objections to the BAI regulations. See also C. J. Marshall, "Veterinary Education," *Journal of the American Veterinary Medical Association* 65 (1924): 28–38, esp. pp. 33 and 37.

84. Fish, "The Veterinary Outlook from a Teacher's Viewpoint," p. 9.

85. Howard Preston, *Automobile Age Atlanta: The Making of a Southern Metropolis 1900–1935* (Athens: University of Georgia Press, 1979), p. 40; *Boyd's Business Directory* (Philadelphia, PA: 1900), pp. 1557–1558; *Polk's-Boyd's Philadelphia Directory, 1925* (Philadelphia, PA: R. L. Polk, 1925), "Classified Section," p. 1261, "Veterinarians." Census data confirmed this conclusion. Philadelphia, for example, counted 109 veterinarians in 1900, 99 in 1910, and 120 in 1920. For data on this and other cities, see Department of Commerce and Labor, Bureau of the Census, *Special Reports, Twelfth Census, 1900: Occupations* (Washington, DC: Government Printing Office, 1904), table 43 (p. 672); Department of Commerce, Bureau of the Census, *Thirteenth Census, 1910, Volume IV: Population, Occupation Statistics* (Washington, DC: Government

Printing Office, 1914), table III (p. 193); Department of Commerce, Bureau of the Census, *Fourteenth Census, 1920, Volume IV: Population, Occupations* (Washington, DC: Government Printing Office, 1923), table 19 (p. 218).

86. Fish, "The Veterinary Outlook from a Teacher's Viewpoint," see p. 2 for quotes. The case of Philadelphia is instructive here. Dr. Howard Felton's Equine-Canine Infirmary offered small-animal care as early as 1895 (see *Boyd's Business Directory* [Philadelphia, PA, 1895], pp. 1432–1433). John J. Maher's Veterinary Hospital for Horses, Dogs, and Small Animals in Philadelphia became a pet hospital over a period of 10 years beginning in 1905. By 1915, Maher's attached farm for boarding horses was no longer mentioned as part of the dog and cat hospital. See *Boyd's Business Directory* (Philadelphia, PA, 1900), pp. 1557–1558; *Boyd's Business Directory* (Philadelphia, PA, 1905), pp. 1884–1887; *Boyd's Business Directory* (Philadelphia, 1910), pp. 2452–2455; *Boyd's Business Directory* (Philadelphia, PA, 1915), pp. 2112–2113. This topic is explored more thoroughly in Chapter Five of this volume.

87. Simon J. J. Harger, "Veterinary Practice in Cities," in University of Pennsylvania, Faculty of the School of Veterinary Medicine, *The Veterinary Profession: Its Relation to the Health and Wealth of the Nation* (Philadelphia, PA: Avil Printing, 1897), p. 15; Fish, "The Veterinary Outlook from a Teacher's Viewpoint," p. 2.

88. George H. Hart, "The Relation of Veterinary Extension Work to the Practicing Veterinarian, the Livestock Industry and the Public Health," *Journal of the American Veterinary Medical Association* 59 (April 1921): 43–50, see pp. 44–45.

89. A. F. Schalk, "Undeveloped Resources in Veterinary Sanitation and Hygiene," *Journal of the American Veterinary Medical Association* 59 (July 1921): 410–433, see p. 411.

90. For a description of factors leading to the agricultural depression, and contemporary hopes for a swift reversal of it, see Henry C. Wallace, "The Turn of the Tide in Agriculture," *Audubon Advocate* (Sept. 7, 1922), pp. 143–144, 161.

91. John L. Shover, *First Majority, Last Minority: The Transforming of Rural Life in America* (DeKalb: Northern Illinois University Press, 1976), p. 242.

92. Fish, "The Veterinary Outlook from a Teacher's Viewpoint," p. 6; Shideler, *Farm Crisis, 1919–1923*, introduction.

93. Fish, "The Veterinary Outlook from a Teacher's Viewpoint," p. 6.

94. S. J. Schilling, "Some of the Newer Problems in Veterinary Science," *Journal of the American Veterinary Medical Association* 69 (April–September 1926): 569–578, quote at p. 578.

95. Of course, the veterinary profession had not worked very hard before World War I to develop companion-animal practice as an important professional activity. Veterinarians simply did not perceive a professional obligation toward animals that were valued sentimentally. This was a philosophical position with economic consequences.

96. For more on veterinarians' professional activities, see J. F. Smithcors, in *The American Veterinary Profession: Its Background and Development* (Ames: Iowa State University Press, 1963) and Bert W. Bierer, in *A Short History of Veterinary Medicine in America* (East Lansing: Michigan State University Press, 1955).

97. On the development of late nineteenth-century scientific institutions, including agricultural experiment stations, see Charles E. Rosenberg, *No Other Gods: On Science and Social Thought* (Baltimore, MD: Johns Hopkins University Press, 1997), chaps. 8–12.

Chapter 3. The Value of Animal Health for Human Health

1. George W. Goler, "Municipal Milk Work in Rochester," *Charities and the Commons* 16 (August 4, 1906): 484–487, quotes at p. 487. A large literature exists on the link between milk and infant and child mortality in the United States. Recent useful studies include Jacqueline H. Wolf, *Don't Kill Your Baby: Public Health and the Decline of Breastfeeding in the 19th and 20th Centuries* (Columbus: Ohio State University Press, 2001); Richard A. Meckel, *Save the Babies: American Public Health Reform and the Prevention of Infant Mortality, 1850–1929* (Baltimore, MD: Johns Hopkins University Press, 1990); and Rima Apple, *Mothers and Medicine: A Social History of Infant Feeding, 1890–1950* (Madison: University of Wisconsin, 1987).

2. "Farmers Talk About Cattle and Trolley; Emphatically Opposed to a Change in the Tuberculosis Inspection Law," *Philadelphia Inquirer,* December 7, 1897, p. 16. New Jersey dairymen supplied milk to the city of Philadelphia, which sought to impose veterinary inspection for tuberculosis on their cows in 1897.

3. Ilyse C. Barkan, "Industry Invites Regulation: The Passage of the Pure Food and Drug Act of 1906," *American Journal of Public Health* 75 (1985): 18–26. For a general introduction to consumer interest in controlling the products available to them, see R. W. Fox and T. J. J. Lears, eds., *The Culture of Consumption: Critical Essays in American History 1880–1980* (New York: Pantheon Books, 1980).

4. Ernest Wende, "City Milk Routes and Their Relation to Infectious Diseases," *Journal of the American Medical Association* 34 (1900): 150; Henry E. Alvord and Raymond A. Pearson, *The Milk Supply of Two Hundred Cities and Towns, USDA Bureau of Animal Industry Bulletin No. 46* (Washington, DC: Government Printing Office, 1903), pp. 101, 107, 110.

5. William Cronon, *Nature's Metropolis: Chicago and the Great West* (New York: W. W. Norton, 1991).

6. Alvord and Pearson, *Milk Supply,* pp. 53, 106, 144; *City of St. Paul v. Peck,* 81 N.W. 389, 78 Minn. 497 (1900).

7. Alvord and Pearson, *Milk Supply,* p. 26; for an example of a court challenge to the "two-cow" rule, see *Pierce v. City of Aurora,* 81 Ill. App. 670 (1899).

8. For examples of prosecutions of milk adulterers, see *Guilder v. State,* 26 Ohio Cir. Ct. R. 221 (1904); *Lansing v. State,* 102 N.W. 254, 73 Neb. 124 (1905); *Wiegand v. District of Columbia,* 22 App. D.C. 559 (1903); *Isenhour v. State,* 62 N.E. 40, 157 Ind. 613, 97 Am. St. Rep. 228 (1901); and the case of Honolulu, Hawaii, in Alvord and Pearson, *Milk Supply,* p. 61.

9. Nathan Straus, "How to Reduce Infant Mortality," March 22, 1897, in Lina Gutherz Straus, *Disease in Milk: The Remedy, Pasteurization. The Life Work of Nathan Straus* (New York: E. P. Dutton, 1917), p. 187. Sanitary recommendations for cows, their surroundings, and milk handlers from this time period are outlined in Sarah

D. Belcher, *Clean Milk* (New York: Orange Judd, 1907), chaps. 1–12. This study also argues that bacteriological testing is necessary for milk. The author, a physician, was employed by the Rockefeller Institute for Medical Research and the research laboratory of the New York City Department of Health.

10. Barbara Gutmann Rosenkrantz, "The Trouble with Bovine Tuberculosis," *Bulletin of the History of Medicine* 59 (1985): 155–175; Charles E. North, "Milk and Its Relation to Public Health," in Mazyck P. Ravenel, ed., *A Half Century of Public Health* (New York: American Public Health Association, 1921), pp. 236–289; Edwin Henry Shorter, "A Summary of Milk Regulations in the United States," *Journal of the American Public Health Association* 11 (1911): 847–856.

11. Alvord and Pearson, *Milk Supply,* pp. 26–29, 57–59, 137–138, 48–49; Belcher, *Clean Milk,* p. 124; C. Hampson Jones, "The Present Needs of the Milk Supply of Baltimore," *Charities and the Commons* 16 (August 4, 1906): 499–504 (this issue of *Charities* was partially devoted to an examination of efforts to clean up the municipal milk supply in the United States). For descriptions of the laboratory tests commonly applied to milk through the early twentieth century, see Louis A. Klein, *Principles and Practice of Milk Hygiene* (Philadelphia, PA: Lippincott, 1917) (Klein was a veterinarian); and Clarence Henry Eckles, Willes Barnes Combs, and Harold Macy, *Milk and Milk Products, Prepared for the Use of Agricultural College Students* (New York: McGraw-Hill, 1929), chaps. 7, 13, and appendices.

12. Goler, "Municipal Milk Work in Rochester," pp. 484–487. For the history of the certified milk movement, see Manfred J. Wasserman, "Henry L. Coit and the Certified Milk Movement in the Development of Modern Pediatrics," *Bulletin of the History of Medicine* 46 (1972): 379–390.

13. Samuel C. Prescott, "The Production of Clean Milk from a Practical Standpoint," *Charities and the Commons* 16 (August 4, 1906): 488–491, quoted at p. 491. The first bacterial counts of milk in the United States were made in Boston in 1892. Standardized methods for counting bacteria in milk were not available until 1908 and this probably delayed recognition of its importance; see Charles E. North, "Milk and Its Relation to Public Health," in Ravenel, *Half Century of Public Health,* p. 283; James O. Jordan, "Boston's Campaign for Clean Milk," *Journal of the American Medical Association* 49 (1907): 1082–1087.

14. "Farmers Talk About Cattle and Trolley," *Philadelphia Inquirer,* December 7, 1897, p. 16. The tuberculin test involved injecting inactivated bacteria into the cow and scrutinizing the animal over several hours for a reaction, which was indicated by a rise in body temperature. For an example of a court case in which a city's right to demand the tuberculin test was upheld, see *State v. Nelson,* 68 N.W. 1066, 66 Minn. 166, 34 L.R.A. 318, 61 Am. St. Rep. 399 (1896).

15. *John Quincy Adams v. City of Milwaukee and Gerhard A. Bading,* 33 S. Ct. 610, 612, decided May 12, 1913, by the U.S. Supreme Court. D. H. Udall to V. A. Moore, April 2, 1902, box "D. H. Udall—Miscellaneous Speeches and Correspondence," Flower-Sprecher Archives, New York State Veterinary College at Cornell University, Ithaca, NY. Before the wide utilization of the tuberculin test, veterinary leaders had based the profession's claim to public health work on dairy sanitation and proper

feeding and breeding of milk cows. See William Herbert Lowe, "The Relation of Veterinary Medicine to the Public Health," *American Veterinary Review* 24 (1901): 589–592.

16. See 228 U.S. 572, 580 for appellate court rulings; and *Adams v. City of Milwaukee*, 33 S. Ct. 610, 612.

17. *Adams v. City of Milwaukee*, 33 S. Ct. 610.

18. Goler, "Municipal Milk Work in Rochester," p. 487; Edward F. Brush, "How to Produce Milk for Infant Feeding," *Journal of the American Medical Association* 43 (November 5, 1904): 1385; Louis A. Klein (referred to as "Dr. Kline"), response to "The Necessity for the Training of Full Time Milk Inspectors," *Proceedings of the 11th, 13th, 14th, 15th, and 16th Annual Conferences of the American Association of Medical Milk Commissions* (Chicago: American Association of Medical Milk Commissions, 1922), p. 493.

19. "Washington Views on Tuberculosis," *Breeder's Gazette* 71 (1917): 73. Bovine tuberculosis eradication efforts increased at the same time that pasteurization was replacing certified milk and decreasing the chance of tuberculosis transmission.

20. "Washington Views on Tuberculosis," p. 73; Virginia C. Meredith, "A Tuberculosis Demonstration," *Breeder's Gazette* 71 (1917): 1289; Leonard Pearson and M. P. Ravenel, *Tuberculosis of Cattle and the Pennsylvania Plan* (Harrisburg, PA: Wm. Stanley Ray, State Printer of Pennsylvania, 1901), see farmer survey numbers 163, 215, and 230, pp. 217, 222, 223; M. H. Reynolds, "Area Testing for Tuberculosis," *Breeder's Gazette* 79 (1921): 598; H. Lyon, "York State Attacking Tuberculosis," *Breeder's Gazette* 79 (1921): 728.

21. Sporadic protests against tuberculin testing occurred well into the 1930s, however; in October 1931, the Iowa National Guard was mobilized to protect veterinarians and restrain farmers armed with "pitchforks, mud and eggs of ancient vintage." See "Tuberculin Test War About Over," *Journal of the American Veterinary Medical Association* 79 (July–December 1931): 427; "Hostilities Continued," *Journal of the American Veterinary Medical Association* 79 (July–December 1931): 581.

22. Wolf, *Don't Kill Your Baby*, chap. 2; Wasserman, "Henry L. Coit and the Certified Milk Movement," pp. 379–390; Jones, "The Present Needs of the Milk Supply of Baltimore," p. 499; Samuel Hopkins Adams, "Rochester's Pure Milk Campaign," *McClure's Magazine* 29 (1907): 142–149. Smaller cities also ran pure milk campaigns; see C. W. M. Brown, "Certified Milk in Small Cities," *Journal of the American Medical Association* 48 (1907): 587–588.

23. H. Wirt Steele, "The Milk Campaign in Maryland," *Charities and the Commons* 16 (August 4, 1906): 474–478, quote at p. 475.

24. Suellen Hoy, *Chasing Dirt: The American Pursuit of Cleanliness* (New York: Oxford University Press, 1995); Belcher, *Clean Milk*, pp. 74–75; Straus, *Disease in Milk*, pp. 180, 198. For an example of municipal milk regulation at this time, see "An Ordinance relating to the maintenance of dairies, and the production, keeping, transportation, sale and distribution, of milk," Series 1909, Supervisor's Bill No. 29, Ordinance No. 88, City and County of Denver, July 16, 1909.

25. Barkan, "Industry Invites Regulation," pp. 18–26; E. R. Larned, "Unclean

Milk, Bovine Tuberculosis and the Tuberculin Test—Their Relation to the Public Health," *Boston Medical and Surgical Journal* 149 (1903): 563–567; Susan D. Jones, "Animal Value, Veterinary Medicine, and the Domestic Animal Economy in the United States, 1890–1930," Ph.D. diss., University of Pennsylvania, Philadelphia, PA, 1997, pp. 148–149; Rosenkrantz, "The Trouble with Bovine Tuberculosis," pp. 170–171; John Duffy, *The Sanitarians: A History of American Public Health* (Urbana: University of Illinois Press, 1990), p. 198.

26. A. S. Wheeler, "Municipal Meat Inspection," in University of Pennsylvania, Faculty of the School of Veterinary Medicine, *The Veterinary Profession: Its Relation to the Health and Wealth of the Nation* (Philadelphia, PA: Avil Printing, 1897), p. 24; *Philadelphia Inquirer,* December 7, 1897, editorial page; *American Veterinary Review* 21 (April 1897 to March 1898): 713; Rima D. Apple, *Mothers and Medicine: A Social History of Infant Feeding, 1890–1950* (Madison: University of Wisconsin Press, 1987), pp. 98–99.

27. Most of this narrative has been taken from Caroline Bartlett Crane's 1909 speech, "Interest in Meat Inspection," box 22, folder 7, Caroline Bartlett Crane Papers, A-92, Archives and Regional History Collections, Western Michigan University, Kalamazoo, Michigan (hereafter Crane Papers); see pp. 1–5. While Crane's devotion to the cause of improved meat inspection was unusual, she nonetheless was only the most visible of thousands of middle-class women, many belonging to municipal branches of the General Federation of Women's Clubs, who were active advocates of pure food reform. Crane was also interested in dairies and pure milk; see Paul U. Kellogg, "Interesting People: Caroline Bartlett Crane," *American Magazine* 69 (December 1906): 172–174; Hanson Booth, "Cleaning Up the American City: How Mrs. Caroline Bartlett Crane Does It," *American Magazine* 76 (September 1913): 45–48.

28. Crane recollected that in the year 1904 alone, the states of Indiana and South Dakota, and the cities of Elkhart and Madison, IN, and Grand Rapids, MI, enacted statutes and ordinances through the efforts of women's clubs advised directly by her. Crane, "Interest in Meat Inspection," pp. 1–19. The other key to Crane's success was her address to the United Master Butchers of America in 1905, which was published with "kindest comments" in *The Butchers and Packers Gazette.* Women in other locales could use the reprints to buttress their arguments in favor of regulation. "When a local butcher can be shown that the United Master Butchers of America (of which he probably is a member through his local branch) is favorable to the movement, it inevitably softens his opposition" (p. 24).

29. Crane, "Interest in Meat Inspection," pp. 2, 3, 10; Daniel E. Salmon, "The Relation of Bovine Tuberculosis to the Public Health," address delivered to the American Public Health Association, September 16, 1901, Buffalo, NY (Salmon was the chairman of the committee on animal diseases and animal food). The address was later published as *USDA, Bureau of Animal Industry Bulletin No. 33* (Washington, DC: Government Printing Office, 1901).

30. Crane, "Interest in Meat Inspection," p. 12; Lowe, "The Relation of Veterinary Medicine to the Public Health," pp. 589, 591. Veterinarians were involved in

local inspection efforts throughout the United States, but especially in towns hosting universities with veterinary programs. See, for example, J. Dennis McGuire and James E. Hansen, *Chiron's Time* (Fort Collins, CO: College of Veterinary Medicine and Biomedical Sciences, 1983), p. 31.

31. "Veterinary Legislation," *American Veterinary Review* 18 (1894): 724–725; *American Veterinary Review* 19 (1895): 104–105; *American Veterinary Review* 19 (1895–96): 677.

32. Pearson was 28 years old when he was appointed state veterinarian. See Ray Thompson, *After 1883: One Hundred Years of Organized Veterinary Medicine in Pennsylvania* (Philadelphia: W. B. Saunders, 1982), pp. 153–155; quotes from Faculty of the University of Pennsylvania, *Leonard Pearson, In Memoriam* (Philadelphia, PA: University of Pennsylvania, 1909), pp. 38–39. Pearson's famous tuberculin test on this occasion demonstrated the impartiality of tuberculosis toward aristocratic as well as more humble animals. The cattle that tested positive were valued at $10,000 and were the property of Joseph Gillingham, president of the board of managers of the University of Pennsylvania Veterinary Hospital, where Pearson worked. Pearson may have hesitated before recommending the slaughter of all the positive reactors, but he was vindicated when postmortem results showed the presence of tuberculosis organisms and lesions in the condemned cattle. Another famous episode of pedigreed, well-cared-for cattle proving tuberculous occurred with the testing of J. P. Morgan's herd in August 1909; one-third of the valuable animals had to be destroyed (see Straus, *Disease in Milk,* p. 55).

33. Rosenkrantz, "The Trouble with Bovine Tuberculosis," pp. 155–175; editorials in the *American Veterinary Review* 25 (September 1901): 476–480; "Moore on Koch," undated news clipping, reproduced in Ellis P. Leonard, compiler, *A Collection of Historical Papers and Memorabilia Assembled from the Archives of the Flower and Olin Libraries: Veranus A. Moore,* vol. I (Ithaca, NY: Flower-Sprecher Veterinary Library, 1983), p. 7. The bacteriological argument hinged on whether the bovine and human tubercle bacilli were "of the same species"; the practical arguments included whether the bovine bacillus was transmissible to and capable of causing disease in humans, and the amount of public health resources that should be utilized to combat it. The bovine bacillus, *Mycobacterium tuberculosis bovis,* is now known to be transmissible to humans through milk and undercooked meat and to cause a sometimes severe intestinal infection and skeletal deformity. The more familiar form of pulmonary and systemic tuberculosis, transmitted between humans, is caused by *Mycobacterium tuberculosis.*

34. Roscoe R. Bell and W. R. Ellis, editorial, *American Veterinary Review* 28 (1904–05): 718.

35. See Pearson and Ravenel, *Tuberculosis of Cattle,* for the specifications of the Pennsylvania Plan.

36. Pearson and Ravenel, *Tuberculosis of Cattle,* pp. 11, 12, 158, quote at p. 158.

37. Pearson and Ravenel, *Tuberculosis of Cattle,* pp. 150, 181, 185. For more details on this and other state bovine tuberculosis eradication plans, see Jones, "Animal Value, Veterinary Medicine, and the Domestic Animal Economy," chap. 2.

38. Pearson and Ravenel, *Tuberculosis of Cattle*, pp. 195–197.

39. Bell and Ellis, editorial, 718–719; Pearson and Ravenel, *Tuberculosis of Cattle*, pp. 178, 195; Alonzo Taylor, in *Leonard Pearson, In Memoriam*, p. 39; Louis A. Klein, "Pioneer Work in Tuberculosis Control," *Journal of the American Veterinary Medical Association* n.s. 11 (1921): 455. The Pennsylvania Plan actually more closely resembled that of Denmark, and Pearson and Ravenel reproduced Bernhardt Bang's essay "Tuberculosis of Cattle and Its Suppression in Denmark" as an appendix to *Tuberculosis of Cattle.*

40. D. E. Salmon, "The Federal Meat Inspection," in USDA, *Yearbook of the Department of Agriculture, 1894* (Washington, DC: Government Printing Office, 1895), pp. 73–74; "Report of the Secretary, 1897," p. 19; "Report of the Secretary," USDA, *Yearbook of the Department of Agriculture, 1905* (Washington, DC: Government Printing Office, 1906), p. 30; "Report of the Secretary," USDA, *Yearbook of the Department of Agriculture, 1906* (Washington, DC: Government Printing Office, 1907), p. 25.

41. In 1892, a total of 9,351,396 beef quarters were inspected, of which 8,160,625 were for the domestic interstate trade. See Salmon, "The Federal Meat Inspection," p. 68.

42. For a detailed description of inspection and disposition procedures, see Rudolf Alexander Clemen, *The American Livestock and Meat Industry* (New York: Ronald Press, 1923), pp. 331–339.

43. Jimmy M. Skaggs, *Prime Cut: Livestock Raising and Meatpacking in the United States, 1607–1983* (College Station: Texas A&M University Press, 1986), pp. 3, 215.

44. Salmon, "The Federal Meat Inspection," pp. 67–77.

45. Vivian Wiser, Larry Mark, and H. Graham Purchase, *100 Years of Animal Health, 1884–1984* (Beltsville, MD: Associates of the National Agricultural Library, 1987), p. 172.

46. Wiser et al., *100 Years of Animal Health*, p. ix; Bert W. Bierer, *American Veterinary History* (mimeographed, © 1940 by the author; reproduced by Carl Olson, 1980), p. 212.

47. D. E. Salmon, "U.S. Meat Inspection," *American Veterinary Review* 26 (April 1895): 31.

48. Harry Thurston Peck, *Twenty Years of the Republic, 1885–1905* (New York: Dodd, Mead, 1920), p. 622. For descriptions of the Beef Trust's activities, see the "Beef Trust Case," *Swift & Co. v. United States*, 196 U.S. 395 ([1902–]1905) and Clemen, *The American Meat and Livestock Industry*, 752–757. The Chicago packers had been under investigation since the 1880s for similar monopolizing arrangements, as described in U.S. Senate, *Testimony Taken by the Select Committee on the Transportation and Sale of Meat Products*, 51st Congress, 1st session, 1889–1890, S. R. 829, Serial 2705 (also known as the "Vest Report"). For a contemporary commentary on the "embalmed beef" scandal, see the New York *World*, December 25 and 26, 1898.

49. Samuel Hopkins Adams, "The Subtle Poisons," *Collier's* 36 (1905): 16–18; Adolphe Smith, "Chicago: The Dark and Insanitary Premises Used for Slaughtering Cattle and Hogs—the Government Inspection," *Lancet* (1905, part I, no. 1): 121–123; Smith, "Chicago: Tuberculosis Among the Stockyard Workers," *Lancet* (1905,

part I, no. 1): 183–185; Smith, "Unhealthy Work in the Stockyards," *Lancet* (1905, part I, no. 1): 358–360; Crane, "Interest in Meat Inspection," pp. 25–26; Charles Edward Russell, "The Greatest Trust in the World," *Everybody's Magazine* 12 (1905): 147–156, 291–300, 503–516, 643–654; 13 (1905): 56–66, 217–227; Russell, *The Greatest Trust in the World* (New York: Ridgway-Thayer, 1905); Upton Sinclair, "The Jungle," *Appeal to Reason* nos. 482–518, February 25, 1905, to November 4, 1905; Sinclair, *The Jungle* (New York: Doubleday, Page, 1906). For more on the 1904 strike and the Beef Anti-Trust case, see Clemen, *American Meat and Livestock Industry*, pp. 699–703, 753, 754. For the packing industry's response, see "Is Chicago Meat Clean?" *Collier's* (April 22, 1905): 14.

50. James A. Wilson to Honorable H. C. Adams, August 24, 1905, in Private Book 12, box 4, RS 9/1/11, James A. ("Tama Jim") Wilson Papers, Iowa State University Archives, Ames, IA (hereafter Wilson Papers). Wilson to C. D. Boardman, August 22, 1905, and Wilson to D. D. Colbrook, September 5, 1905, in Private Book 12, box 4, Wilson Papers; Wilson to Honorable W. B. Allison, September 26, 1905, in Private Book 12, box 4, Wilson Papers. Veterinary historians have commonly attributed Salmon's resignation to "being thrown to the lions in the wake of sensational charges concerning the meat inspection" [J. F. Smithcors, *The American Veterinary Profession: Its Background and Development* (Ames: Iowa State University Press, 1963), p. 495; Smithcors, *The Veterinarian in America, 1625–1975* (Goleta, CA: American Veterinary Publications, 1975), p. 93; Bierer, *American Veterinary History*, p. 106 and epilogue]. However, Salmon resigned in November, before the worst of the meat controversy hit the USDA, and Secretary Wilson bore the brunt of its fury. Salmon spent the next 5 years starting a veterinary school in Uruguay, then returned to the United States, worked on the Anaconda Copper Mine case, and died in Butte, Montana, on August 30, 1914. James A. Wilson to Honorable J. P. Dollinger, September 29, 1905, and Wilson to Murdo Mackenzie (American Stock Growers' Association), October 16, 1905, in Private Book 12, box 4, Wilson Papers.

51. James A. Wilson to William E. Curtis, January 24, 1906, in Private Book 12, box 4, Wilson Papers. For BAI employees, the climax of departmental concern came with a confidential order printed in April asking all employees to "answer at once, without evasion," with a full accounting of "stocks, shares, certificates, bonds, rights, and interests" owned by employees and their families as well as any gifts or fees received from companies engaged in meat production. See memorandum dated June 5, 1906, on letterhead of the U.S. Army Purchasing Commissary, in D. Arthur Hughes, "1906 Meat Inspection Scrapbook," box 3, RS 21/7/73, Clarence H. Pals Papers, Iowa State University Archives, Ames, IA (hereafter Pals Papers). Upton Sinclair, *The Autobiography of Upton Sinclair* (New York: Harcourt, Brace, and World, 1962), p. 126; Christine Scriabine, "The Writing of *The Jungle*," *Chicago History* 10 (Spring 1981): 26–37; Christine Scriabine, "Upton Sinclair: Witness to History," Ph.D. diss., Brown University, Providence, RI, 1973, esp. pp. 9–12, 55–70. USDA, Amendment No. 5 to BAI Order No. 125, released February 14, 1906, as "Rules and Regulations for the Inspection of Live Stock and Their Products." Caroline Crane labeled it a "valentine"; see Crane, "Interest in Meat Inspection," p. 26. J. Ogden Armour,

"The Private Freight Car System," *Saturday Evening Post* 178, January 6, 1906, p. 1; "The Private Car Controversy," *Saturday Evening Post* 178, January 20, 1906, p. 1; "The Packers and the Cattlemen," *Saturday Evening Post* 178, February 10, 1906, p. 12; "The Packers and the People," *Saturday Evening Post* 178, March 3, 1906, p. 1; "The Packers and the Future," *Saturday Evening Post* 178, March 24, 1906, p. 10, and March 31, 1906, p. 13.

52. W. K. Jaques, "A Picture of Meat Inspection," *World's Work* (May 1906): 7491–7514. Doubleday, Page, published both this journal and Sinclair's *The Jungle*. D. Arthur Hughes, "1906 Meat Inspection Scrapbook," box 3, Pals Papers, pp. 8–9.

53. James A. Wilson to Honorable Redfield Proctor, May 22, 1906; for evidence of presidential pressure, see James A. Wilson to Theodore Roosevelt, April 11, 1906, in which Wilson disclaims responsibility for a leak of information to the press. Both letters in Private Book 12, box 4, Wilson Papers.

54. "The 'Knockers' and the Knocked," *Breeder's Gazette* (May 16, 1906): 1035.

55. Salmon, "Federal Meat Inspection," p. 77.

56. Caroline Bartlett Crane, "What Is Happening to American Meat Inspection?" (1909), pp. 10–11, A-92, box 22, folder 8, Crane Papers; Alonzo Melvin, "The Federal Meat Inspection Service," *USDA, Bureau of Animal Industry Circular* (Washington, DC: Government Printing Office, February 28, 1908). The BAI decision to salvage muscle tissue and other parts of carcasses from tubercular animals was apparently based on similar policies of meat inspection in European countries and the recommendations of researchers there (including E. I. E. Nocard and Robert Koch). For a good synopsis of research on this question, see Pearson and Ravenel, *Tuberculosis of Cattle*, chap. 4 and pp. 73–79.

57. See BAI orders, S.&R.A. 107, 1914, and "Disposition of Tuberculous Carcasses," May, 1909, in box 3, first binder (unprocessed), Pals Papers.

58. The two most important legislative years were 1912 and 1921. House Resolution 512 (62nd Congress, 2nd session), sponsored by Wisconsin Representative John M. Nelson in April 1912 at the instigation of Caroline Crane and Albert Leffingwell, criticized the BAI, questioned the validity of passing as inspected parts of carcasses from diseased animals, and demanded an investigation and further legislation of meat inspection. It failed only after a massive lobbying effort by meat packers, veterinarians, and livestock breeders. Some veterinary leaders suspected that the American Medical Association was sponsoring Crane's investigative work, "endeavoring to place the veterinarians in as bad a light as possible" after being "foiled in their endeavors to put over the Owen bill." See letter to D. Arthur Hughes from Joseph Hughes, Chicago Veterinary College, May 10, 1912, attached to D. A. Hughes scrapbook, Pals Papers, p. 140 (news clippings and other materials relating to the 1912 episode are affixed to pp. 134–141). The issue of healthy meat did not disappear from public discourse, however, and legislation allowing consumer complaint and financial reparation from meat-packing companies was finally passed in 1921 (H.R. 6320, 67th Congress, revised in 1926 and 1935).

59. North, "Milk and Its Relation to Public Health," pp. 272, 277.

Chapter 4. The Value in Numbers

1. Harvey Levenstein, *Paradox of Plenty: A Social History of Eating in Modern America* (New York: Oxford University Press, 1993).

2. The process of "separating the [raising and] killing of an animal from the eating of its flesh" had been continuing at least since the Civil War, as William Cronon discusses in *Nature's Metropolis: Chicago and the Great West* (New York: W. W. Norton, 1991), pp. 207–259, 264.

3. For a statement of veterinarians' attitudes on the importance of food production, see Arthur D. Goldhaft, *The Golden Egg* (New York: Horizon Press, 1957), pp. 260, 277, 279–280.

4. For general accounts of these two disease eradication campaigns, see J. F. Smithcors, *The American Veterinary Profession: Its Background and Development* (Ames: Iowa State University Press, 1963), pp. 451–457; O. H. V. Stalheim, *The Winning of Animal Health: 100 Years of Veterinary Medicine* (Ames: Iowa State University Press, 1994), chap. 5. The Bureau of Animal Industry's (BAI) annual reports, years 1908–1953, chart the agency's descriptions of its work in disease eradication. For popular support of the bovine tuberculosis eradication campaign and cheap milk, see Lewis Edwin Theiss, "Why Not Clean Up the Milk in the Country?" *Outlook* 133 (January 24, 1923): 176–179.

5. Report of the Committee on Education, *Journal of the American Veterinary Medical Association* 79 (November 1931): 692–693; Goldhaft, *Golden Egg*, pp. 133–140; John R. Mohler, "Fifty Fruitful Years in Veterinary Science," *Journal of the American Veterinary Medical Association* 87 (1935): 64–73. For personal stories of veterinarians' problems during the depression, see Stalheim, *Winning of Animal Health*, p. 6.

6. See Chapter Three in this volume for a discussion of veterinarians' earlier work on bovine tuberculosis.

7. J. A. Kiernan, "Tuberculosis and Our Livestock Industry," *Journal of the American Veterinary Medical Association* 54 (October 1918–March 1919): 107–109; J. A. Kiernan, "Tuberculosis Eradication," *Journal of the American Veterinary Medical Association* 54 (October 1918–March 1919): 604; E. V. Moore, "The Eradication of Bovine Tuberculosis from Isolated Centers," *Veterinary Medicine* 43 (1948): 17–20; J. A. Kiernan and L. B. Ernest, "The Toll of Tuberculosis in Livestock," *USDA Yearbook of Agriculture*, 1919 (Washington, DC: U.S. Government Printing Office, 1920), pp. 277–288. Kiernan's eulogist characterized his campaign as one based on "economic grounds" rather than a public health mandate. Certainly this is the picture Kiernan presented in livestock journals, the *Yearbooks of Agriculture*, and the veterinary journals. Nonetheless, the initial appropriations for the campaign, and continuing funding for the tuberculosis eradication division, depended upon the acknowledgment of the disease as a human health issue. As historian J. F. Smithcors wrote, "the economic, social, and purely medical aspects of most public health problems can hardly be separated into neat compartments" (*American Veterinary Profession*, p. 470).

8. Kiernan and Ernest, "The Toll of Tuberculosis," pp. 287–288. In securing appropriations, Kiernan received assistance from the dairy farming lobby; see A. J. Glover to Hon. William M. Jardine, December 26, 1925 (Bureau of Animal Industry, Central Correspondence, 1913–1953), box 5, file 1.014, RG 17, NARA, College Park, MD. See also records of annual appropriations and supporting documents in Bureau of Animal Industry, Central Correspondence, 1913–1953, box 4, file 1.014, RG 17, NARA. By the time the campaign began in 1917, dairy journals were largely in favor of tuberculosis eradication. See, for example: Editorial, "The Sick Cow," *Kimball's Dairy Farmer* 6 (September 1, 1904): 4; Query Letter, "Applying the Tuberculin Test," *Hoard's Dairyman* 46 (October 31, 1913): 391; "The Application of the Tuberculin Test" (summary of Pennsylvania Experiment Station Bulletin 123), *Hoard's Dairyman* 46 (January 9, 1914): 718; and Malcom H. Gardner, letter, "Bovine Tuberculosis," *Hoard's Dairyman* 46 (January 9, 1914): 724. For veterinarians' viewpoint, see the annual reports of the special committee on tuberculosis, delivered at the American Veterinary Medical Association's convention between 1919 and 1940, and published in the *Journal of the American Veterinary Medical Association*.

9. Mohler, "Fifty Fruitful Years in Veterinary Science," pp. 64–73; R. R. Dykstra, "Veterinary Medicine in Kansas," *Bulletin of the Kansas Veterinary Medical Association* 9, no. 1 (1952): 13; see also the discussion of hog cholera and bovine tuberculosis campaigns in B. W. Kingrey, "Farm Animal Practice in the United States," *Journal of the American Veterinary Medical Association* 169, no. 1 (1976): 56–60. Kingrey argues that the hog cholera campaign in particular had its disadvantages for the image of the veterinary profession, owing to the behavior of some practitioners who functioned only as hog vaccinators and "did not even own a stethoscope" (p. 58). Conversely, historian O. H. V. Stalheim has characterized the hog cholera campaign as one of the most important factors in "professionalizing" veterinary medicine (see Stalheim, *The Winning of Animal Health*, pp. 176–183).

10. Philip M. Teigen has used the word "zoogeography" to describe distributions of domesticated animals; see Teigen and Sheryl A. Blair, "The Massachusetts Veterinary Profession, 1882–1904," *Historical Journal of Massachusetts* 25 (1997): 63–73 and Teigen, "Historical Zoogeography of the Urban Horse in the U.S., 1860–1920," paper presented at the 25th Symposium of the International Committee for the History of Technology, August 18–22, 1998, Lisbon, Portugal; for further discussion of this term, see Chris Philo and Chris Wilbert, eds., *Animal Spaces, Beastly Places: New Geographies of Human-Animal Relations* (London: Routledge, 2000), pp. 4,8.

11. "Report of the Special Committee on Meat Hygiene," *Journal of the American Veterinary Medical Association* 87 (July–December 1935), p. 433; "Thirty-One States Now Accredited," *Journal of the American Veterinary Medical Association* 87 (July–December 1935): 615–616.

12. [Louis A. Merillat] Editorial, "Is Veterinary Practice Keeping Pace with the General Progress of Veterinary Science?" *Veterinary Medicine* 16 (July 1921): 38–40. See discussion of this in Smithcors, *American Veterinary Profession*, pp. 558–562.

J. V. Lacroix, editor of *North American Veterinarian,* opposed Merillat in this viewpoint. See, for example, his editorial "Veterinarians and the Public Health," *North American Veterinarian* 4 (January 1923): 3–4.

13. Dean Cooper, "Relationship Between the Veterinarian and the County Agent," *Journal of the American Veterinary Medical Association* 9 (October 1919– March 1920): 153–158; "The Hog Cholera Situation," editorial, *Journal of the American Veterinary Medical Association* 70 (December 1926): 279–281. For more on the origins of agricultural extension work, see "Cooperative Agricultural Extension Work," *United States Department of Agriculture Circular No. 47,* May 3, 1915 (Washington, DC: Government Printing Office, 1915).

14. J. S. Koen, E. E. Chast, G. E. Totten, E. B. Bennett, and Charles S. Chase, "Report of the Special Committee on Meat Hygiene," *Journal of the American Veterinary Medical Association* 87 (1935): 431–442, quote at p. 433; O. H. V. Stalheim, *Veterinary Conversations with Mid-Twentieth-Century Leaders* (Ames: Iowa State University Press, 1996), p. xx. See also editorial, "Should We Tackle Hog Cholera Eradication Now?" *Journal of the American Veterinary Medical Association* 121 (December 1952): 489–491.

15. John Parascandola, ed., *The History of Antibiotics: A Symposium* (Madison, WI: American Institute of the History of Pharmacy, 1980); H. C. Smith, "Proper Use of Penicillin," *Veterinary Medicine* 42 (1947): 85; William Arthur Hagan and Dorsey William Bruner, *The Infectious Diseases of Domestic Animals, with Special Reference to Etiology, Diagnosis, and Biologic Therapy,* 2nd ed. (Ithaca, NY: Comstock, 1951), pp. 91–100. The first edition of this book, published in 1943, did not yet contain a special section on antimicrobial agents. C. M. Stowe, P. B. Hammond, A. L. Aronson, F. H. Kriewaldt, "A Survey of Some of the Pharmacological Properties of Four Sulfonamides in Dairy Cattle," *Cornell Veterinarian* 47 (October 1957): 469–479; "it is freely predicted . . . " quoted from *Veterinary Medicine* 42 (1947): 83; K. D. Downham and G. J. Christie, "Treatment of Mastitis in Dairy Cows with Penicillin," *Veterinary Medicine* 42 (1947): 98–99.

16. Advertisements described: Abbott's Crystalline Penicillin-G, *Veterinary Medicine* 42 (November 1947): xxxi; Commercial Solvents Crystalline Penicillin-C.S.C., *Veterinary Medicine* 42 (April 1947): x and December 1947, p. xvii; Lederle "Veticillin," advertised as a bovine mastitis treatment, *Veterinary Medicine* 42 (November 1947): xxi; Parke-Davis Penicillin-G, *Veterinary Medicine* 42 (December 1947): iii.

17. R. L. Rudy, "Dosage of Penicillin," *Veterinary Medicine* 42 (1947): 326; "Penicillin for Large Animals," *Veterinary Medicine* 42 (1947): 445; "Use of Penicillin in Leptospirosis," *Veterinary Medicine* 42 (1947): 434; Smith, "Proper Use of Penicillin," p. 85; Herminie B. Kitchen and Selman A. Waksman, "Streptomycin and Neomycin in Veterinary Medicine," *Journal of the American Veterinary Medical Association* 127 (July–December 1955): 261–274; F. W. Schofield, cited in "Notes from Purdue Conference," *Journal of the American Veterinary Medical Association* 125 (December 1954): 447; Lederle Laboratories, *Aureomycin: Its Application in Veterinary Medicine* (New York: Lederle Laboratories, 1952).

18. Myron G. Fincher, "The Accomplishments of William A. Hagan During 25

Years as Dean of the New York State Veterinary College," *Cornell Veterinarian* 48 (1958): 231–238.

19. Ronald Colston, "In 1946, Good Management Will Pay Dividends," *Poultry Tribune* 52 (January 1946): 13, 35–37; Clinton P. Anderson, cited in Fred Bailey, "Washington Bulletin," *Poultry Tribune* 52 (January 1946): 42; "Importance of an Animal Industry," *Veterinary Medicine* 43 (1948): 135; "Looking Ahead," editorial, *Poultry Tribune* 52 (April 1946): 5; O. A. Hanke, "Our Answer to Poultry-less and Egg-less Thursdays," *Poultry Tribune* 53 (November 1947): 62; editorial, "Feed's Function in a War Economy," *Feed Age* 1 (January 1951): 19. Worldwide food shortages were exacerbated by poor growing seasons in 1945, 1946 and 1947. Cereal crop shortages eased somewhat in 1948.

20. John G. Hardenburgh, "Whence and Whither?" *Veterinary Medicine* 43 (1948): 277–280; "The Veterinary School, University of Minnesota," *Veterinary Medicine* 43 (1948): 312.

21. Margaret Rossiter, "The Organization of the Agricultural Sciences," in Alexandra Oleson and John Voss, *The Organization of Knowledge in Modern America, 1860–1920* (Baltimore, MD: Johns Hopkins University Press, 1979), pp. 211–248; W. H. Peters, "President's Address," *Journal of Animal Science* 1 (1942): 52–56; see also Charles Rosenberg's analysis of agricultural experiment stations in *No Other Gods: On Science and American Social Thought* (Baltimore, MD: Johns Hopkins University Press, 1997), pp. 153–210. Most of the BAI's research work was published in periodic bulletins and in the *Journal of Agricultural Research*.

22. Steven Stoll, *The Fruits of Natural Advantage: Making the Industrial Countryside in California* (Berkeley: University of California Press, 1998); Jon Lauck, *American Agriculture and the Problem of Monopoly: The Political Economy of Grain Belt Farming, 1953–1980* (Lincoln: University of Nebraska Press, 2000).

23. The *Good Housekeeping* "Susan" series recommended foods and preparation methods; chicken and turkey accounted for 20% of the articles in this series in 1947. The poultry industry credited "general educational sources," including newspaper and magazine articles, with teaching the consumer to buy poultry products; see O. A. Hanke, "Poultry Industry 'Sizemeter,'" *Poultry Tribune* 53 (February 1947): 14.

24. Page Smith and Charles Daniel, *The Chicken Book* (Boston: Little, Brown, 1975), pp. 262–263; True D. Morse, undersecretary of agriculture, cited in editorial, "Poultry Industry Progress," *Journal of the American Veterinary Medical Association* 125 (October 1954): 332; T. C. Byerly, "Why Hens Are Laying 30 Percent More Eggs," *Poultry Tribune* 54 (April 1948): 11; J. Nickerson, "Chicken in the Pot," *New York Times Magazine* (January 18, 1948), p. 28. Joanna Swabe has described, in the European context, the development of intensive poultry husbandry as the prototype for other species; see Swabe, *Animals, Disease, and Human Society: Human-Animal Relations and the Rise of Veterinary Medicine* (London: Routledge, 1999), pp. 128–135.

25. Louis M. Hurd, *Practical Poultry Farming* (New York: Macmillan, 1928), p. 241.

26. L. C. Norris, "How to Figure Poultry Vitamin Needs: More and More Poultry Raisers Are Leaving This Task to Feed Manufacturers," *Poultry Tribune* 53 (Feb-

ruary 1947): 14, 15, 32; E. I. Robertson, "Why Poultry Feeds Are Better," *Poultry Tribune* 54 (November 1948): 6, 37; L. N. Gilmore, "How Poultry Feeds Are Changing," *Poultry Tribune* 55 (February 1949): 11, 20, 21. Based on the need for "scientific diets," feed manufacturing for livestock grew tremendously after the war; see E. I. Robertson, "Today's Research in Tomorrow's Feed," *Feed Age* 1 (January 1951): 23–26. Popular articles also described the "improved" chicken and its surroundings: see R. C. Punnett, "The Rise of the Poultry Industry," *The Nineteenth Century* 107 (April 1930): 535–547; Arnold Nicholson, "More White Meat for You," *Saturday Evening Post* 220 (August 9, 1947): 12 (this article describes the "Chicken of Tomorrow" contest, a nationwide poultry-breeding competition); Richard Thruelsen, "Hens Take No Holidays," *Saturday Evening Post* 219 (October 12, 1946): 22–23. This argument is not intended to diminish the importance of advertising, changing consumption patterns, and other sociocultural considerations; poultry producers were well aware of their importance at this time. See, for example, W. D. Termohlen and M. D. Atkin, "Coming—Revolution in Meat!" *Poultry Tribune* 53 (April 1947): 10, 11, 42; G. B. Wood, "What's Ahead for 1947," *Everybody's Poultry Magazine* 52 (January 1947): 10, 11, 39.

27. Hurd, *Practical Poultry Farming*, pp. 166–177; Thruelsen, "Hens Take No Holidays," p. 23; John Vandervort and F. H. Leuschner, "Six Suggestions to Keep Income High," *Poultry Tribune* 53 (February 1947): 18.

28. Herbert L. Schaller, "We're Winning the War Against Pullorum," *Poultry Tribune* 55 (February 1949): 13, 50, 51; Vandervort and Leuschner, "Six Suggestions to Keep Income High," 18; W. R. Hinshaw, comment on debeaking chickens, "Veterinary Problems in Poultry Production," *Veterinary Medicine* 43 (1948): 472; Arthur D. Goldhaft, *The Golden Egg* (New York: Horizon Press, 1957), pp. 170–177. Notable disease outbreaks immediately after the war also included avian pneumoencephalitis (Newcastle disease). After 1943, the causative virus was known to be passed from chickens to human beings, contributing to concerns over its rapid spread in the United States. See R. P. Hanson and C. A. Brandly, "Newcastle Disease," *Annals of the New York Academy of Sciences* 70, art. 3 (June 3, 1958): 585–597.

29. See, for example, Lederle advertisement, "Sulfaguanidine Stops Coccidiosis," *Poultry Tribune* 52 (June 1946): 9; Karl Seeger, "New Sulfa KO's 'Coxy,'" *Poultry Tribune* 54 (March 1948): 16, 52.

30. Stephen Gordeuk and Marion Learned, "Observations on the Chemotherapy of Coccidiosis," *Veterinary Medicine* 42 (1947): 223–228; Karl C. Seeger, "Sulfas Stop Coccidiosis," *Poultry Tribune* 53 (March 1947): 30–31; Ashton C. Cuckler, Walther H. Ott, and Donald E. Fogg, "Factors in the Evaluation of Coccidiostats in Poultry," *Cornell Veterinarian* 47 (1957): 400–412; "Duatok" advertisement, Lederle Laboratories, *Poultry Tribune* 52 (January 1946): 32; "Coccidiosis Preventive Suggested," *Poultry Tribune* 52 (January 1946): 56; W. R. Hinshaw, "Veterinary Problems in Poultry Production," *Veterinary Medicine* 42 (1947): 413; "The Costly Toll of Animal Diseases," *Feed Age* 3 (February 1953): 40–44.

31. See, for example, Frank Galer, "Fowl Pox Is a Pickpocket—Fail It by Vaccination," *Poultry Tribune* 52 (May 1946): 3, 17–19; F. R. Beaudette, "Be Safe—Use Vac-

cines Properly," *Poultry Tribune* 55 (June 1949): 7, 13, 14; J. P. Delaplane, "Is It Coryza—or Just Plain Roup?" *Poultry Tribune* 52 (Sept 1946): 9. Delaplane, himself a veterinarian at the Rhode Island experiment station, did not specifically advise the consultation of a veterinarian in his article. In 1947, however, *Everybody's Poultry Magazine* inaugurated a veterinary department, with questions answered by veterinarian Morris Povar; the October issue also included a feature article explaining the poultry industry's need for veterinary service.

32. F. C. Tucker, "Poultry Industry Needs Veterinary Service," *Veterinary Medicine* 42 (1947): 24.

33. Goldhaft, *Golden Egg*, pp. 256–257, 260–261, 277–278.

34. Cuckler et al., "Factors in the Evaluation," pp. 401, 408–412.

35. "Section on Poultry," meeting program, *Journal of the American Veterinary Medical Association* 125 (July 1954): 23; editorial, "Science and Feed Manufacturing," *Journal of the American Veterinary Medical Association* 125 (July 1954): 76; "Looking Ahead," editorial, *Poultry Tribune* 52 (July–August 1946): 5. In 1947, *Veterinary Medicine* published a group of articles designed to inform veterinarians about poultry diseases in its February issue (vol. 42, pp. 68–73); it also printed a regular feature on veterinary problems in poultry production. The feed industry also called for "closer cooperation between the feed manufacturer and the veterinarian"; see editorial, "The Feed Industry and the Veterinarian," *Feed Age* 1 (October 1951): 11.

36. "Newcastle Vaccine Developed by USDA," *Everybody's Poultry Magazine* 52 (April 1947): 32; "Greater Recognition Sought by Poultry Industry," *Veterinary Medicine* 42 (1947): 90; "Poultry Meat Inspection," *Veterinary Medicine* 42 (1947): 2. Although not federally mandated, inspection was growing in popularity, owing to municipalities' implementation of inspection laws specifically requiring veterinary inspection. See "USDA Poultry Meat Inspection," *Veterinary Medicine* 42 (1947): 49, an adaptation of H. A. Weckler's interview broadcast on Chicago radio station WGN, November 16, 1946.

37. E. A. Schmoker, "Clinical Reports: Cow, General Delivery, Woodinville," *Veterinary Medicine* 42 (February 1947): 79–81.

38. O. A. Hanke, "Views from 6,000 Miles" (editorial), *Poultry Tribune* 57 (March 1951): 66, and Hanke, "Wanted: $3000 for an Egg IQ Program," *Poultry Tribune* 57 (April 1951): 58; Hanson and Brandly, "Newcastle Disease," p. 595; editorial, "Disease Control by Insanitation," *Journal of the American Veterinary Medical Association* 127 (October 1955): 370–371.

39. Thomas H. Jukes, "The History of the 'Antibiotic Growth Effect,'" *Federation Proceedings* 37, no. 11 (1977): 2514–2518; Thomas H. Jukes, "Dr. Best, Insulin, and Molecular Evolution," *Canadian Journal of Biochemistry* 57 (1979): 455–456; Thomas H. Jukes, letter to the editor, *Clinical Toxicology* 14, no. 3 (1979): 319–322. Jukes became the scientist most identified with the antibiotic growth effect, although he was careful to point out that "the results flowed from many laboratories in 1949."

40. J. E. Briggs and W. M. Beeson, "Effects of Vitamin B_{12}, Aureomycin, Streptomycin, or Dried Whey Factor Supplements on the Growth and Fattening of Weanling Pigs," *Journal of Animal Science* 11 (1952): 103–111.

41. Kitchen and Waksman, "Streptomycin and Neomycin," pp. 263–264; Jukes, "The History of the 'Antibiotic Growth Effect,'" p. 2516.

42. W. S. Gordon and J. H. Taylor, "Antibiotics as Supplements to the Rations of Farm Livestock," *Veterinary Record* 65 (1953): 838; O. A. Hanke, "Let's All Face the Facts," *Poultry Tribune* 57 (October 1951): 54; J. H. Florea, "What's the Outlook for '52?" (pp. 7, 15) and Fred H. Masterson, "Ohio Joins the Broiler Boom" (pp. 8, 31) *Poultry Tribune* 57 (December 1951); E. D. Parnell, "Feed's Effect on the Quality of Poultry Products," *Feed Age* 3 (February 1953): 26–27.

43. R. N. Arsdall and H. C. Gilliam, in L. P. Schertz et al. (eds.), *Another Revolution in United States Farming?* (Washington, DC: United States Dept of Agriculture, 1979), pp. 190–256; Dr. Jack Schmitz, personal communication, May 31, 2000; H. Ernest Bechtel, "Life Accelerators," *Feed Age* 1 (September 1951): 20–24; Joseph C. Blight, "The Story of Antibiotics in Swine Nutrition," *Feed Age* 1 (October 1951): 12–15, 50; Jerry Thompson, "Antibiotic Supplements Lower Feeding Costs," *Feed Age* 1 (November 1951): 39–40; "Antibiotic Roundup," *Feed Age* 3 (November 1953): 29–38.

44. Vivian Wiser, Larry Mark, and H. Graham Purchase, *100 Years of Animal Health, 1884–1984* (Beltsville, MD: Associates of the National Agricultural Library, 1987).

45. *Modern Veterinary Practice* Staff Report, "Can Dairy Practice Survive?" *Modern Veterinary Practice* 50 (February 1969): 31–34.

46. Editorial [J. F. Smithcors], "1968 + 16 = 1984," *Modern Veterinary Practice* 49 (January 1968): 35–38, quote at p. 38.

47. Editorial, "The New Farmer," *Journal of the American Veterinary Medical Association* 147, no.7 (1965): 758 (originally appeared in the June 1965 issue of *Agri Dynamics*); editorial, "Whither Farm Animal Practice?" *Journal of the American Veterinary Medical Association* 147, no. 5 (1965): 519.

48. Editorial Viewpoints, "Cattlemen Recognize Veterinarians' Role," *Journal of the American Veterinary Medical Association* 147, no. 1 (1965): 72; O. M. Radostits and D. C. Blood, *Herd Health: A Textbook of Health and Production Management of Agricultural Animals* (Philadelphia, PA: W. B. Saunders, 1985), p. 4.

49. Swabe, *Animals, Disease and Human Society,* pp. 118–119; Smithcors, *The Veterinarian in America,* p. 93; Cronon, *Nature's Metropolis,* pp. 255–257.

50. *Modern Veterinary Practice* Staff Report, "Can Dairy Practice Survive?" 31–34, quote at p. 32.

51. Michael W. Fox, *Farm Animals: Husbandry, Behavior, and Veterinary Practice (Viewpoints of a Critic)* (Baltimore, MD: University Park Press, 1984). Fox, a proponent of animal welfare, argued that animals in confined situations developed "learned helplessness" as a result of their inability to "mobilize their usual adaptive responses" (see appendix B).

52. *Modern Veterinary Practice* Staff Report, "Can Dairy Practice Survive?" *Modern Veterinary Practice* 49 (February 1969): 31–34; *Modern Veterinary Practice* Staff Report, "Hog Cholera Eradication: The Producers Speak Out," *Modern Veterinary Practice* 51 (August 1970): 19–22; Allen F. Schauffler, "Is Livestock Practice Going to the Dogs?" *Veterinary Medicine—Small Animal Clinician* 64 (October 1969): 854–858.

53. *Modern Veterinary Practice* Staff Report, "Hog Cholera Eradication: What's the Impact?" *Modern Veterinary Practice* 51 (July 1970): 21–24. This survey of sixteen veterinarians engaged in general rural practice found that five of the nine who had taken specific steps to ensure their futures as veterinarians had become small-animal, or pet, practitioners. Schauffler, "Is Livestock Practice Going to the Dogs?" pp. 854–858.

Chapter 5. Pricing the Priceless Pet

1. For the purposes of this chapter, "small animal," "companion animal," and "pet" will be considered interchangeable terms for animal populations composed largely of dogs, cats, and other small pet animals. Although space limits the species considered, other animals could and certainly did become pets in the same manner described here for dogs and cats.

2. Center for Information Management, *U.S. Pet Ownership and Demographics Sourcebook* (Schaumburg, IL: American Veterinary Medical Association, 1997), pp. 1, 3, 105, 11, 117, 121. See also *PFI Fact Sheet* (Washington, DC: Pet Food Institute, 1998).

3. Keith Thomas, *Man and the Natural World: A History of the Modern Sensibility* (New York: Pantheon, 1983). For more on the development of Victorian compassion for companion animals, see James Turner, *Reckoning with the Beast: Animals, Pain, and Humanity in the Victorian Mind* (Baltimore, MD: Johns Hopkins University Press, 1980).

4. Max J. Herzberg, ed., *The Terhune Omnibus* (Cleveland, OH: World Publishing, 1945, © 1937), pp. 22, 23, 28.

5. Martin Green, "Fine Traits Shown by the Dogs of War," from the New York *Evening World*, excerpted in *Literary Digest* 59 (December 14, 1918): 47–48. For a more detailed discussion, see Susan D. Jones, "Animal Value, Veterinary Medicine, and the Domestic Animal Economy in the United States, 1890–1930," Ph.D. diss., University of Pennsylvania, Philadelphia, PA, 1997, chap. 5.

6. Allan A. ("Scotty") Allan, "The Bravest Dogs in the World," *Sunset Magazine* 54 (April 1925): 8–9, 63, 75–76.

7. For a detailed account of these events, see Kenneth A. Ungermann, *The Race to Nome* (New York: Harper & Row, 1963). Although Ungermann's book is somewhat dramatized and is classified as juvenile literature, its value lies in the careful primary research, including interviews with participants, that he conducted.

8. Catharine Brody, "Dog Stars and Horse Heroes," *Saturday Evening Post* 197 (February 14, 1925): 17, 96, quote at p. 101; John B. Kennedy, "Trailing the Dog Star," *Collier's* 78 (August 21, 1926): 7, 28. The German shepherd breed was especially popular for movie work; see Edward Tenner, "The Road to Rin Tin Tin: Social Construction of the German Shepherd Dog," paper presented to Shelby Cullom Davis Center Seminar, Princeton University, September 19, 1997.

9. "Dog Food: NRA Watches to See That Pups Get Decent Meals," *Newsweek* 4 (Oct. 6, 1934): 30.

10. Gladys Anderson Emerson, "Some Nutrition Studies with Dogs," *North*

American Veterinarian 17 (September 1936): 29; American Animal Hospital Association, *Fifty Years of Educational Excellence*, p. 5; *Newsweek*, "Dog Food," p. 30; editorial, "Senators Discuss Canned Dog Foods," *Journal of the American Veterinary Medical Association* 88 (January–June 1936): 692–693; editorial, "Change in Regulations Covering Dog Foods," *Journal of the American Veterinary Medical Association* 89 (July–December, 1936): 632.

11. The expansion of national advertising to include a wider range of products and causes during the 1920s has been documented by Daniel Pope, *The Making of Modern Advertising* (New York: Basic Books, 1983). For an examination of how cultural values were incorporated into advertising in America, see Roland Marchand, *Advertising and the American Dream: Making Way for Modernity, 1920–1940* (Berkeley: University of California Press, 1985).

12. "Rin-Tin-Tin's Reward," advertisement for Ken-L Ration, *Country Life* 54 (May 1928): 25; and "The Worker and His Food," advertisement for Ken-L Meal, *American Kennel Club Gazette* 49 (June 1, 1932): 3.

13. See, for example, "Lassie Thrives on Red Heart," advertisement, *Dog World* 34 (June 1949): 19; "Lassie Recommends Red Heart 3-Flavor Dog Food," advertisement, *Dog World* 34 (August 1949): 17.

14. "The Worker and His Food," p. 3.

15. "Lassie Thrives on Red Heart," p. 19; "Lassie Recommends Red Heart 3-Flavor Dog Food," see p. 17 for quote.

16. *Advertising Age*, October 4, 1937, p. 29. This advertisement was cited in Marchand, *Advertising and the American Dream*, p. 358.

17. Advertisement for Spratt's biscuits, *Journal of the American Veterinary Medical Association* 86 (June 1935): 13.

18. J. C. Flynn, "Canine Practice," *American Veterinary Review* 42 (1912–1913): 176–183, quotes at pp. 176, 182.

19. Joanna Swabe has also made this argument; see *Animals, Disease, and Human Society*, p. 180.

20. "U. of P. Clinic Report for 1923," *Journal of the American Veterinary Medical Association* 65 (1924): 38; Charles Gatchell, "Hospital for Dumb Animals," *Technical World Magazine* 19 (July 1913): 726–728. Hospitals run by humane organizations provided care free of charge, and this was undoubtedly one of their attractions. Nonetheless, they promoted the idea that pets deserved veterinary care.

21. "Section on Small Animals," *Journal of the American Veterinary Medical Association* 77 (July–December 1930): 667; Willard C. Haselbush, *Mark Morris: Veterinarian* ([U.S. : s.n.]: R. R. Donnelly, 1984). See also Fred D. Patterson, Jr., "Internal Parasites of the Canine and Feline," *Journal of the American Veterinary Medical Association* 69 (April–September 1926): 157–166; William M. Bell, "Abdominal Surgery in the Canine and Feline," *Journal of the American Veterinary Medical Association* 69 (April–September 1926): 167–172; F. F. Parker, "Cesarean Section in the Bitch," *Journal of the American Veterinary Medical Association* 77 (January–June 1930): 781.

22. J. C. Flynn, "The Needs of the General Practitioner in Small Animal Practice," *Journal of the American Veterinary Medical Association* 69 (April–Sept. 1926):

724–728, quote at p. 724; H. Schreck, "Hospital Arrangement and Management," *Journal of the American Veterinary Medical Association* 69 (April–Sept. 1926): 48–52; A. A. Feist, "As a Client Looks at Our Profession," *Journal of the American Veterinary Medical Association* 69 (April–Sept. 1926): 463–466; F. F. Parker, "Cesarean Section in the Bitch," *Journal of the American Veterinary Medical Association* 76 (Jan.–June 1930): 781–784, quote at p. 781.

23. For more on the importance of laboratory science to veterinary training, see Jones, "Animal Value, Veterinary Medicine, and the Domestic Animal Economy," chaps. 4 and 5.

24. Although antivivisectionists in the United States had largely focused on physicians, the Anglo-American movement owed its beginnings in part to its founder Frances Cobbe's horror at accounts of vivisections carried out on horses at the veterinary school in Alfort, France. For a table of 1920s federal antivivisection legislation, see Jones, "Animal Value, Veterinary Medicine, and the Domestic Animal Economy," chap. 5.

25. For more on the development of early twentieth-century veterinary research, see Jones, "Animal Value, Veterinary Medicine, and the Domestic Animal Economy," chap. 4. See also editorial, "Why an AVMA Research Fund?" *Journal of the American Veterinary Medical Association* 108 (February 1946): 107–108; "Committee Report: Legislation," *Journal of the American Veterinary Medical Association* 109 (November 1946): 420.

26. On the District of Columbia dog pound as a source for Bureau of Animal Industry (BAI) research dogs, see B. H. Rawl to W. C. Fowler, April 25, 1919; B. H. Rawl to W. C. Fowler, August 24, 1920; W. C. Fowler to B. H. Rawl, August 27, 1920; all in RG 17, file 1.208, box 33, Bureau of Animal Industry, file 1.092, box 133, NARA. Other veterinary researchers at the BAI interested in black tongue and pellagra included Jacob Shillinger and Leigh T. Giltner [see G. A. Wheeler, "Black Tongue in Dogs and Its Relation to Pellagra in the Human," *Journal of the American Veterinary Medical Association* 77 (July–December 1930): 62–72]. Haselbush, *Mark Morris*, pp. 73–76.

27. Maurice Hall's correspondence with opponents of animal experimentation is in RG 17, box 327, file 2.299, NARA. The transcripts of the 1930 hearings can be found as (1) U.S. House of Representatives, "Hearing Before the Committee on the District of Columbia, on H.R. 7884, a Bill to Prohibit Experiments on Living Dogs in the District of Columbia" (Washington, DC: U.S. Government Printing Office, 1931) (this record is hereafter referred to as H.R. 7884 Hearing) and (2) U.S. Senate, "Hearings Before the Committee on the District of Columbia, on S. 4497, a Bill to Prohibit Experiments upon Living Dogs in the District of Columbia," June 6, 10, 11, 12, 1930 (Washington, DC: U.S. Government Printing Office, 1930) (this record is hereafter referred to as S. 4497 Hearing).

28. "Statement of Dr. Maurice C. Hall," attached to letter dated June 12, 1930, from Hall to Watson Davis, RG 17, box 327, file 2.299, NARA; S. 4497 Hearing, pp. 90, 98; Buckingham quote at p. 125.

29. Susan E. Lederer, *Subjected to Science: Human Experimentation in America Be-*

fore the Second World War (Baltimore, MD: Johns Hopkins University Press, 1995), esp. chaps. 1, 2, and 4; Richard D. French, *Antivivisection and Medical Science in Victorian Society* (Princeton, NJ: Princeton University Press, 1975), pp. 272–273.

30. S. 4497 Hearing, pp. 33–35.

31. S. 4497 Hearing, pp. 123, 126.

32. Maurice C. Hall to J. M. Cattell, May 13, 1931, RG 17, box 327, file 2.299, NARA.

33. Maurice C. Hall to J. V. Lacroix, June 23, 1930, RG 17, box 327, file 2.299, NARA. Hall appeared to be shaken by this episode; he wrote to several colleagues about it and in his 1932 article "The Prevention of Cruelty" described some antivivisectionists as "sadists who conceal under an outward love of animals a cruelty towards mankind" (p. 214); Hall, "The Prevention of Cruelty and the Work of a Great Humane Society," *Scientific Monthly* 34 (March 1932): 211–221.

34. Maurice C. Hall to J. M. Cattell, May 13, 1931; "The Prevention of Cruelty" is the article that Hall submitted in 1931.

35. Hall, "The Prevention of Cruelty," pp. 211–216, quote at p. 221. When mentioning "human cruelty" here, Hall was referring to bureau efforts on behalf of animal transportation and slaughterhouse regulations.

36. This argument was not unique to Hall; see, for example, Pierre A. Fish to Hon. Gale Stalker, December 13, 1930, reproduced in Ellis Leonard, ed., *Pierre A. Fish Correspondence,* bound, p. 254, Flower-Sprecher Veterinary Library, New York State College of Veterinary Medicine, Cornell University. Excerpts from Hall's article were reprinted as late as 1948 in veterinary journals; see "A Great Humane Program," *Veterinary Medicine* 43 (January 1948): 6.

37. J. Elliott Crawford, "The Hospitalization of Small Animals," *Journal of the American Veterinary Medical Association* 74 (December–June 1928–1929): 743–751, quote at p. 743. This was the first of a series of articles published in the journal between 1928 and 1931 advising veterinarians how to build and run a hospital for companion animals. See also W. E. Frink, "Small-Animal Hospitalization," *Journal of the American Veterinary Medical Association* 74 (December–June 1928–1929): 713–720; Horst Schreck, "The Veterinary Hospital, with Special Reference to Accomodations [*sic*] for Small Animals," *Journal of the American Veterinary Medical Association* 79 (July–December 1931): 170–179, 332–348, 586–602, 728–751.

38. A. A. Feist, "As a Client or the Public Looks at Our Profession," *Journal of the American Veterinary Medical Association* 69 (April–September 1926): 463–466, quote at p. 465. Veterinary educators also believed that women could not physically handle larger animals. See Phyllis Hickney Larsen, ed., *Our History of Women in Veterinary Medicine* (Madison, WI: Omnipress, for the Association for Women Veterinarians, 1997), pp. 8–9.

39. Veterinary practitioner quoted in American Animal Hospital Association, *Fifty Years of Educational Excellence,* p. 5.

40. Pierre A. Fish, "Why the Veterinary Profession?" *Cornell Veterinarian* 20 (July 1930): 302.

41. N. S. Mayo et al., "Report of the Committee on Education," *Journal of the American Veterinary Medical Association* 79 (November 1931): 669–702, see pp. 697–

698. Veterinarians (both rural and urban) polled spent 38, 24, and 19% of their professional time on cattle, pets, and horses, respectively.

42. H. P. Rusk to Earl C. Smith, October 29, 1938, RG 17, file 1.092, box 133, NARA.

43. John D. Blaisdell, "A Frightful—But Not Necessarily Fatal—Madness: Rabies in Eighteenth-Century England and English North America," Ph.D. diss., Iowa State University, Ames, IA, 1995; Blaisdell, "The American Veterinary Community's Reception of Pasteur's Work on Rabies," *Agricultural History* 70 (Summer 1996): 503–524. Bert Hansen has shown the importance of rabies in establishing medical research on diseases as a major focus of popular interest in the late nineteenth-century United States; see Hansen, "America's First Medical Breakthrough: How Popular Excitement About a French Rabies Cure in 1885 Raised New Expectations for Medical Progress," *American Historical Review* 103 (April 1998): 373–418.

44. Thomas G. Hull, "The Spread and Control of Rabies," *Journal of the American Veterinary Medical Association* 74 (December 1928–June 1929): 1047–1051; B. J. Killham, "Rabies Control in Michigan," *Journal of the American Veterinary Medical Association* 74 (December 1928–June 1929): 183–192. Killham's paper was read at the American Veterinary Medical Association's meeting on August 8, 1928, and the discussion following it clearly indicated practitioners' differing beliefs about which formulations of vaccine were effective.

45. Killham, "Rabies Control in Michigan," p. 187; M. F. Barnes and A. N. Metcalfe, "Investigations of Canine Diseases, with Special Reference to Rabies" *Journal of the American Veterinary Medical Association* 76 (January–June 1930): 34–52; F. H. Brown, "The Control of Rabies in Indiana," *Journal of the American Veterinary Medical Association* 74 (December 1928–June 1929): 178–182; Lester C. Neer, "Rabies Control in Ohio," *Journal of the American Veterinary Medical Association* 78 (January–June 1931): 708–709. For physicians' recommendations as understood by veterinarians, see Julius H. Hess, "Clinical Aspects of Human and Animal Rabies," *Journal of the American Veterinary Medical Association* 76 (January–June 1930): 420–432; David J. Davis, "The Unity of Human and Veterinary Medicine," *Journal of the American Veterinary Medical Association* 86 (January–June 1935): 258; "Rabies: A Plea for Conquest Through Cooperation," *Hygeia* 15 (August 1937): 726–727; editorial, "The Rabies Menace," *Journal of the American Veterinary Medical Association* 91 (July–December 1937): 5–6 (comments on a *Journal of the American Medical Association* article).

46. H.W. Schoening, "Experimental Studies with Killed Canine Rabies Vaccine," *Journal of the American Veterinary Medical Association* 76 (January–June 1930): 25–33; Schoening, "Prophylactic Vaccination of Dogs Against Rabies," *Journal of the American Veterinary Medical Association* 78 (January–June 1931): 703–707; John Reicherl and J. E. Schneider, "Rabies Vaccine Protection Test," *Journal of the American Veterinary Medical Association* 84 (January–June 1934): 752–758. Vaccine efficacies were still controversial among scientists in 1934, however; see M. F. Barnes, A. N. Metcalfe and W. E. Martindale, "Canine Rabies Experimental Vaccination," *Journal of the American Veterinary Medical Association* 84 (January–June 1934): 740–751. By 1936,

however, Parke, Davis & Company was advertising its chloroform-killed vaccine, citing Schoening's work. See, for example, advertisement in *Journal of the American Veterinary Medical Association* 42 (October 1936): advertising supplement p. 5.

47. "Rabies in Illinois and Michigan," *Journal of the American Veterinary Medical Association* 88 (January–June 1936): 731; Robert J. Foster, "Address of the President," *Journal of the American Veterinary Medical Association* 91 (July–December 1937): 271–272; H. M. Kalodner, "Report of Committee on Rabies," *Journal of the American Veterinary Medical Association* 90 (January–June 1937): 408–418; C. E. DeCamp, "The Prophylactic Vaccination of Dogs Against Rabies in a Public Health Program," *Journal of the American Veterinary Medical Association* 91 (July–December 1937): 581–587. DeCamp polled veterinary schools and found only one actively promoting the use of the vaccine in the community; this was probably Auburn (Alabama), which was a partner in a Rockefeller Institute–funded study of prophylactic canine vaccination. The impetus among veterinarians for mandatory vaccination arose largely from the BAI and from local veterinary organizations and pet practitioners.

48. Harry W. Jakeman, "Some Phases of Rabies and Distemper Control," *Journal of the American Veterinary Medical Association* 90 (January–June 1937): 493–500; quote from "Rabies Control Possible If Public Demands It," *Science News Letter* 34 (September 17, 1938): 183; "Rabies Control in New York State," *Veterinary Medicine* 42 (May 1947): 168. For popular educational campaigns, see *Newsweek* article on rabies campaign commented on in Wilson T. Sowder, "The Relationship of the Veterinarians to Public Health Work," *Veterinary Medicine* 43 (March 1948): 102–106; Russell K. Lowry, "The Truth About Rabies," *Better Homes and Gardens* 24 (August 1946): 66, 80, 81; "For Man and Dog," *Time* 54 (September 12, 1949): 56; "Vaccinate Against Rabies," *Science News Letter* 50 (July 20, 1946): 45; "Campaign to Wipe Out Rabies in Dogs," *Science News Letter* 62 (August 2, 1952): 71; and "National Anti-Rabies Effort," *Science News Letter* 62 (July 5, 1952): 13. For standardization of vaccine, see Raymond Fagan, "Rabies Control," *Veterinary Medicine* 43 (July 1948): 295–298.

49. "Canine Distemper," *Journal of the American Veterinary Medical Association* 74 (December 1928–June 1929): 967–968.

50. M. L. Morris, "Further Studies in the Control and Hospitalization of Canine Distemper," *Journal of the American Veterinary Medical Association* 85 (July–December 1934): 39–63; George Watson Little, "Serum Concentration—Living Virus Immunity Against Canine Distemper," *Journal of the American Veterinary Medical Association* 85 (July–December 1934): 577–596; M. L. Morris, "Clinical and Laboratory Studies of the Simultaneous Use of Serum Concentrate (Little) and Living Virus for Immunizing Dogs Against Distemper," *North American Veterinarian* 15 (1934): 32–37. Adolph Eichhorn, a distinguished researcher at Lederle Laboratories, reminded veterinarians that all of the American work had been predicated on that of the English researchers; see A. Eichhorn, "Credit Where Credit Is Due" (letter to the editor), *Journal of the American Veterinary Medical Association* 85 (July–December 1934): 823–824. He referred to P. P. Laidlaw and G. W. Dunkin, "Studies on Dog Distemper III. The Nature of the Virus," *Journal of Comparative Pathology and Ther-*

apeutics 39 (1926): 3, 222; P. P. Laidlaw and G. W. Dunkin, "Report of Field Investigations of Canine Distemper," *Veterinary Medicine* 24 (May 1929): 210–215; and G. W. Dunkin and P. P. Laidlaw, "Some Further Observations on Dog Distemper I. The Durability of the Immunity Following Vaccine and Virus Administration," *Journal of the American Veterinary Medical Association* 78 (July–December 1931): 545–551. The importance of practitioners carrying out research lay in part with its persistence; well into the 1950s, large studies on the field use of vaccines were being reported by private practitioners. See, for example, Edward J. Scanlon and Robert D. Barndt, "Field Use of Avianized Canine Distemper Vaccine," *Journal of the American Veterinary Medical Association* 125 (July–December 1954): 55–61.

51. See, for example, A. S. Schlingman, "Studies on Canine Distemper IV. Immunization of Dogs by Means of Bacterial Products," and commentary following, *Journal of the American Veterinary Medical Association* 83 (July–December 1933): 604–617.

52. See, for example, *Journal of the American Veterinary Medical Association* advertising supplements. Ads for Lederle Laboratories (vol. 87, June 1935, p. 8; vol. 88, January 1936, p. 8); Jen-Sal (vol. 87, December 1935, p. 1); and Pitman-Moore (vol. 87, July 1935, p. 3; vol. 88, May 1936, p. 3; vol. 91, December 1937, p. 3).

53. "Preventive medicine" also described veterinary procedures, including vaccination, used on other species of animals. For a listing of cat-related books and journals with dates of founding, see Susan D. Jones, "Framing Animal Disease: Housecats with Feline Urological Syndrome, Their Owners, and Their Doctors," *Journal of the History of Medicine and Allied Sciences* 52 (April 1997), p. 209, note 16. Estimates of money spent on cat food taken from "Dog and Cat Food Trends," *Journal of the American Veterinary Medical Association* 161 (December 15, 1972): 1678. For vaccine, see, for example, Lederle Laboratories advertisement, *Journal of the American Veterinary Medical Association* 87 (June 1935): p. 8 of advertising supplement.

54. For example, see Pitman-Moore advertisement, *Journal of the American Veterinary Medical Association* 91 (December 1937): p. 3, advertising supplement.

55. D. J. Francisco, letter to the editor, *Journal of the American Veterinary Medical Association* 125 (July–December 1954): 84–85. On the improved rabies vaccine, see "New Rabies Vaccine," *Science News Letter* 57 (May 6, 1950): 279; "Rabies Vaccine for Dogs," *Science Digest* 28 (August 1950): 55; "Improved Agents for Rabies Control," *Journal of the American Veterinary Medical Association* 127 (July–December 1955): 73.

56. Joseph Stetson, letter to the editor, *Scientific American* 192 (May 1955): 2–3; "Cook County Annual Rabies Vaccination Program Announced," *Journal of the American Veterinary Medical Association* 125 (July–December 1954): 191; "Governor Signs State Rabies Law," *Journal of the American Veterinary Medical Association* 131 (July–December 1957): 248.

57. The development of the modern human hospital was probably the greatest influence on animal hospital design. As Charles Rosenberg has shown, human hospitals had matured into their twentieth-century form in the 1920s. See Rosenberg, *The Care of Strangers: The Rise of America's Hospital System* (New York: Basic Books, 1987).

58. J. Elliott Crawford, "The Hospitalization of Small Animals," *Journal of the American Veterinary Medical Association* 74 (December–June 1928–29): 743–751, quote at p. 747; W. E. Frink, "Small-Animal Hospitalization," *Journal of the American Veterinary Medical Association* 77 (July–December 1930): 713–720, quote at p. 713.

59. Frink, "Small-Animal Hospitalization," 717; Horst Schreck, "The Veterinary Hospital: With Special Reference to Accommodations for Small Animals," *Journal of the American Veterinary Medical Association* 79 (July–December 1931): 170–179, 332–348, 586–602, 728–751, quote at p. 596. Besides Schreck's articles, features appearing in *North American Veterinarian* were practitioners' other major source of information.

60. Frink, "Small-Animal Hospitalization," p. 718.

61. See advertisements for "The Veterinary Hospital" in booklet form; for example, *Journal of the American Veterinary Medical Association* 42 (November 1936): advertising supplement p. 10; Horst Schreck, "Reflections on the Planning of Small Animal Hospitals," *Veterinary Medicine* 43 (October 1948): 415–423.

62. *Fifty Years of Educational Excellence and Practice Improvement: AAHA* 50 ([Mishawaka, IN]: American Animal Hospital Association, 1983); quotes from editorial, "Zoning Brochure," *Journal of the American Veterinary Medical Association* 145 (July–December 1964): 1211–1212.

63. Barbara Stein, interview, in Sue Drum and H. Ellen Whiteley, *Women in Veterinary Medicine: Profiles of Success* (Ames: Iowa State University Press, 1991), p. 219; Jones, "Framing Animal Disease," pp. 202–235.

64. "Good for Man and Beast," *Science News Letter* 59 (January 6, 1951): 4; David J. Rothman, *Beginnings Count: The Technological Imperative in American Health Care* (New York: Oxford University Press, 1997).

65. "Better Deal for Dogs," *Newsweek* 37 (January 15, 1951): 58–59; J. C. Furnas, "The Pooches Never Had It So Good," *Saturday Evening Post* 227 (June 18, 1955): 44, 45, 64, 69, 73; Burton J. Rowles, "Your Dog's Next-Best Friend," *Saturday Evening Post* 228 (May 19, 1956): 48, 49, 60, 64, 69.

66. David M. Drenan, "The Growth and Development of Small Animal Practice in the United States," *Journal of the American Veterinary Medical Association* 169 (July 1, 1976): 42–49 (see quotes on pp. 48–49); "Statement of Dr. Maurice C. Hall," June 12, 1930.

67. Feist, "As a Client Looks at Our Profession," p. 464.

68. See Iris M. White, "Business Procedures in a Hospital," *Journal of the American Veterinary Medical Association* 125 (July–December 1954): 70–72, for an example of a discussion of the role of the veterinary wife, who was to "relieve" her husband of "all except medical duties" in the practice.

69. "Meet—Dr. Helen Richt Irwin," *Journal of the American Veterinary Medical Association* 91 (July–December 1937): 259; Urban and Rural Systems Associates, "Exploratory Study of Women in the Health Professions Schools, vol. V: Women in Veterinary Medicine," report prepared for the Women's Action Program, Office of Special Concerns, Office of the Assistant Secretary for Planning and Evaluation, Department of Health, Education, and Welfare (Washington, DC, 1976); Phyllis

Hickney Larsen, ed., *Our History of Women in Veterinary Medicine: Gumption, Grace, Grit and Good Humor* (Madison, WI: Omnipress, for the Association for Women Veterinarians, 1997); Lauralyn J. Brown, "Women in the Veterinary Profession: Yesterday and Today," senior honors thesis, Hampshire College, Amherst, MA, 1987; Susan D. Jones, "Gender and Veterinary Medicine—Global Perspectives," paper presented at World Association for the History of Veterinary Medicine, Annual Congress, Brno, Czech Republic, September 8, 1999.

Chapter 6. Reconciling Use and Humanitarianism

1. Sydney H. Coleman, *Humane Society Leaders in America, with a Sketch of the Early History of the Humane Movement in England* (Albany, NY: American Humane Association, 1924); Francis K. Rowley, *The Humane Idea: A Brief History of Man's Attitude Toward the Other Animals, and of the Development of the Humane Spirit into Organized Societies* (Boston: American Humane Education Society, 1912), p. 52.

2. Roswell McCrea, *The Humane Movement: A Descriptive Survey* (New York: Columbia University Press, 1910), esp. chap. 5; Harriet Ritvo, *The Animal Estate: The English and Other Creatures in the Victorian Age* (Cambridge, MA: Harvard University Press, 1987), pp. 131–132; Keith Thomas, *Man and the Natural World: A History of the Modern Sensibility* (New York: Pantheon Books, 1983).

3. James Turner, *Reckoning with the Beast: Animals, Pain, and Humanity in the Victorian Mind* (Baltimore, MD: Johns Hopkins University Press, 1980).

4. Marian Scholtmeijer, *Animal Victims in Modern Fiction: From Sanctity to Sacrifice* (Toronto: University of Toronto Press, 1993), p. 142.

5. U.S. Department of Commerce and Labor, Census Office, "Slaughtering and Meat Packing," *Census Reports, Volume IX, Twelfth Census of the United States, 1900, Manufactures, Part III* (Washington, DC: Government Printing Offices, 1902), p. 387.

6. U.S. Department of Commerce and Labor, Census Office, *Census Reports, Volume II, Twelfth Census, 1900: Population, Part II* (Washington, DC: Government Printing Office, 1905), pp. 505–507; William Cronon, *Nature's Metropolis: Chicago and the Great West* (New York: W. W. Norton, 1991), chap. 5.

7. Thomas Paine quote from *The Age of Reason, Part I*; for an influential late nineteenth-century edition, see Daniel Conway Moncure, ed., *Writings of Thomas Paine* (New York: Putnam & Sons), p. 83. Moncure was a Unitarian minister interested in human rights and humanitarianism. Benjamin Rush, "An Introductory Lecture to a Course of Lectures . . . upon the Duty and Advantages of Studying the Diseases of Domestic Animals," *Memoirs of the Philadelphia Society for Promoting Agriculture*, vol. I (Philadelphia, PA: Jane Aitken, 1808), pp. xxxxix–lxv, quote at p. lx; Rowley, *The Humane Idea*, pp. 39–41.

8. Rowley, *The Humane Idea*, pp. 5, 6; V[eranus] A. Moore, "The Protection of Domesticated Animals," *Popular Science Monthly* 82 (December 1913): 581.

9. V[eranus] A. Moore, "Some Problems for the Veterinarian," *Journal of the American Veterinary Medical Association* 67 (n.s. 20) (April 1925): 1–9, quote at p. 1.

10. Moore, "Some Problems," p. 1.

11. Rowley, *The Humane Idea*, pp. 58–59.

12. Rowley, *The Humane Idea*, p. 1.

13. Turner, *Reckoning with the Beast*, pp. 29–34.

14. "All About Milk: Our Best Food," pamphlet issued by Metropolitan Life Insurance Company in 1929, p. 11, John W. Hartman Center for Sales, Advertising, and Marketing History, Rare Book, Manuscript, and Special Collection Library, Duke University, Chapel Hill, NC.

15. Jacqueline H. Wolf, *Don't Kill Your Baby: Public Health and the Decline of Breast-feeding in the 19th and 20th Centuries* (Columbus: Ohio State University Press, 2001), pp. 71–72.

16. E. D. Parnell, "Feed's Effect on the Quality of Poultry Products," *Feed Age* 3 (February 1953): 26–27.

17. Cronon, *Nature's Metropolis*, p. 267.

18. Yi-Fu Tuan, *Dominance and Affection: The Making of Pets* (New Haven, CT: Yale University Press, 1984); Ritvo, *The Animal Estate*, p. 40.

19. James Serpell, *In the Company of Animals: A Study of Human-Animal Relationships* (Cambridge: Cambridge University Press, 1986), pp. 19–20.

20. Philip M. Teigen, "Nineteenth-Century Veterinary Medicine as an Urban Profession," *Veterinary Heritage* 23 (May 2000): 1–5.

21. Moore, "The Protection of Domesticated Animals," p. 586.

22. Serpell, *In the Company of Animals*, p. 16, describes the size of English and American investments in pet keeping in the late twentieth century.

23. "Statement of Dr. Maurice C. Hall," attached to letter dated June 12, 1930, from Hall to Watson Davis, RG 17 (Bureau of Animal Industry), box 327, file 2.299, NARA.

24. Ritvo, *The Animal Estate*, p. 162.

25. Letters between W. W. Keen and J. R. Mohler, dated September 1922 to March 1929, RG 17 (Bureau of Animal Industry), box 327, file 2.299, NARA.

26. V[eranus] A. Moore, "Animal Experimentation: The Protection It Affords to Animals Themselves and Its Value in the Livestock Industry of the Country," Defense of Research Pamphlet VI (Chicago: American Medical Association, 1909); "Statement of Dr. Maurice Hall," June 12, 1930; Maurice C. Hall, "The Prevention of Cruelty and the Work of a Great Humane Society," *Scientific Monthly* 34 (March 1932): 216; Maurice Hall's correspondence with opponents of animal experimentation may be found in RG 17 (BAI), box 327, file 2.299, NARA.

27. Susan E. Lederer, "Political Animals: The Shaping of Biomedical Research Literature in Twentieth-Century America," *Isis* 83 (1992): 61–79; Harriet Ritvo, "Plus ça Change: Anti-Vivisection Then and Now," *Science, Technology, and Human Values* 9 (Spring 1984): 57–66.

28. Maurice C. Hall to J. V. Lacroix, June 23, 1930, RG 17 (Bureau of Animal Industry), box 327, file 2.299, NARA.

29. Todd S. Purdum, "Meat Inspections Facing Overhaul, First in 90 Years," *New York Times* (national edition), August 7, 1996, p. 1; Greg Winter, "Illnesses Carried By Food a Problem Despite Advances; 5,000 Such Deaths a Year," *New York Times* (national edition), March 18, 2001, p. 1.

30. "Tainted Food at America's Table," including letter to the editor from Rebecca Goldburg, *New York Times* (national edition), March 25, 2001, section 4, p. 14.

31. "When the Geneticists' Fingers Get in the Food," *New York Times*, February 20, 1994, section 4, p. 14.

32. Scott C. Ratzan, ed., *The Mad Cow Crisis: Health and the Public Good* (New York: New York University Press, 1998); Sheldon Rampton and John Stauber, *Mad Cow USA: Could the Nightmare Happen Here?* (Monroe, ME: Common Courage Press, 1997); Sandra Blakeslee, "On Watch for Any Hint of Mad Cow Disease," *New York Times*, January 30, 2001, p. D3; Donald G. McNeil, Jr., "Epidemic Errors," *New York Times*, February 9, 2001, section 4, p. 1; Geoffrey Cowley, "Cannibals to Cows: The Path of a Deadly Disease," *Newsweek* 138 (March 12, 2001): 53–61.

33. Quotes from Patricia Leigh Brown, "The Warp and Woof of Identity Politics for Pets," *New York Times* (national edition), March 18, 2001, section 4, p. 4; Gerry W. Beyer, "Pet Animals: What Happens When Their Humans Die?" *Santa Clara Law Review* 40, no. 3 (2000): 617–676.

Essay on Sources

Although this book owes much to previous scholarship, it has no direct precedent. Its construction has depended on gleaning source material from disparate places. For historians of the twentieth century who are accustomed to overabundance and one-stop source shopping, the hunt for information about animals can be daunting. Thus, this essay is more of a practical introduction to sources and less of a historiographical discussion than usual.

The intertwined histories of domesticated animals and veterinary medicine in the United States have received little attention from historians. This dearth of scholarship has persisted despite popular interest over the past 30 years in animal welfare and animal rights, food safety, and the roles of companion animals. European authors have produced the most useful recent historical studies of human-animal relationships and animal health care. Two primary examples in English are historian Lise Wilkinson's *Animals and Disease: An Introduction to the History of Comparative Medicine* (Cambridge: Cambridge University Press, 1992) and historian Joanna Swabe's *Animals, Disease, and Human Society: Human-Animal Relations and the Rise of Veterinary Medicine* (London: Routledge, 1999). Swabe takes the *longue durée* approach, with discussions ranging from the archeological evidence of domestication to present-day issues such as bovine spongiform encephalopathy. Wilkinson's work elegantly links the intellectual history of concepts within comparative medicine to the social, political, and professional developments that influenced disease containment. Calvin Schwabe, a veterinary epidemiologist, provided an important prototype with his *Cattle, Priests, and Progress in Medicine* (Minneapolis: University of Minnesota Press, 1978). These books reflect the development of a literature based on relations between humans and animals, grounded in the social and cultural context of time and place.

As such, they have departed significantly from the book-length histories of animal healing that preceded them. In the United States, histories of veterinary medicine have been written largely by veterinarians. They are useful as descriptive narratives or reference manuals to people and events in the history of veterinary medicine; however, they often lack documentation and analysis. Examples include

Bert W. Bierer, *American Veterinary History* (mimeographed; reproduced by Carl Olson, 1980; copyright 1940 by the author); Bierer, *A Short History of Veterinary Medicine in America* (East Lansing: Michigan State University Press, 1955); J. F. Smithcors, *Evolution of the Veterinary Art* (Kansas City, MO: Veterinary Medicine Publishing, 1957); Smithcors, *The American Veterinary Profession: Its Background and Development* (Ames: Iowa State University Press, 1963); and Smithcors, *The Veterinarian in America, 1625–1975* (Goleta, CA: American Veterinary Publications, 1975). Smithcors' work points to important nineteenth-century sources such as agricultural journals and early veterinary advice manuals. A more recent addition to this genre is Robert H. Dunlop and David J. Williams' beautifully illustrated *Veterinary Medicine: An Illustrated History* (St. Louis, MO: Mosby-Yearbook, 1996). Accounts of major events in the history of veterinary medicine include Jeanne N. Logue's account of the Texas cattle fever research, *Beyond the Germ Theory: The Story of Dr. Cooper Curtice* (College Station: Texas A&M University Press, 1995); O. H. V. Stalheim's description of pharmaceutical companies working with veterinarians, *The Winning of Animal Health: 100 Years of Veterinary Medicine* (Ames: Iowa State University Press, 1994); and Bert W. Bierer, *History of Animal Plagues of North America* (printed by the author, copyright 1939, reprinted in Washington, DC: U.S. Department of Agriculture, 1974).

Veterinarians have also written histories of state associations, specialty organizations, and schools. Representative examples include Joseph Arbrura, *Narrative of the Veterinary Profession in California* (San Francisco: published by the author, 1966); Phyllis Hickney Larsen, *Our History of Women in Veterinary Medicine* (Madison, WI: Omnipress, for the Association for Women Veterinarians, 1997); Ellis Pierson Leonard, *A Cornell Heritage: Veterinary Medicine 1868–1908* (1979) and *In the James Law Tradition, 1908–1948* (1982) (Ithaca: Vail-Ballou Press for the New York State College of Veterinary Medicine). Especially with the success of James Herriot's *All Creatures Great and Small* (New York: St. Martin's Press, 1972), numerous American veterinarians have published personal histories. Examples used here include Arthur Goldhaft, *The Golden Egg* (New York: Horizon Press, 1957), and Willard C. Haselbush, *Mark Morris, Veterinarian* ([U.S. : s.n.]: R.R. Donnelley & Sons, 1984). Collections of oral histories include Sue Drum and Ellen Whiteley, *Women in Veterinary Medicine: Profiles of Success* (Ames: Iowa State University Press, 1991), and O. H. V. Stalheim, *Veterinary Conversations with Mid-Twentieth Century Leaders* (Ames: Iowa State University Press, 1996). The best source of articles on the history of veterinary medicine in the United States is the journal of the American Veterinary Medical History Society, *Veterinary Heritage*. There are two histories of the Bureau of Animal Industry that supply some information about its founding and specific projects. Owing to their age, both also function as primary sources. See Ulysses Grant Houck, *The Bureau of Animal Industry of the United States Department of Agriculture: Its Establishment, Achievements, and Current Activities* (Washington, DC: published by the author, 1924) and Fred W. Powell, *The Bureau of Animal Industry: Its History, Activities, and Organizations* (Baltimore, MD: Johns Hopkins Press, 1927).

Important work on the history of veterinary medicine is currently emerging in conference and published papers. This scholarship, archivally based, considers aspects of American veterinary history within a social context. See Philip M. Teigen and Sheryl Blair, "Massachusetts Veterinarians at the End of the Nineteenth Century: A Case Study," *Historical Journal of Massachusetts* 25 (1997): 63–73; and Teigen, "Nineteenth-Century Veterinary Medicine as an Urban Profession," *Veterinary Heritage* 23 (May 2000): 1–6. Of course, this work (including the present book) owes much to the development of the contextual history of (human) medicine and public health over the past 20 years. Historians of veterinary medicine can benefit from the ideas, theories, and methodologies demonstrated in this fast-growing literature. Recommended points of entry include Jacalyn Duffin, *History of Medicine: A Scandalously Short Introduction* (Toronto: University of Toronto, 1999); Charles Rosenberg, *No Other Gods: On Science and American Social Thought* (Baltimore, MD: Johns Hopkins University Press, 1976; rev. and exp. ed., 1997), and *The Care of Strangers: The Rise of America's Hospital System* (New York: Basic Books, 1987); Charles E. Rosenberg and Janet Golden, *Framing Disease: Studies in Cultural History* (New Brunswick, NJ: Rutgers University Press, 1992); John Harley Warner, *Against the Spirit of System: The French Impulse in Nineteenth-Century American Medicine* (Princeton, NJ: Princeton University Press, 1998); and Michael Worboys, *Spreading Germs: Disease Theories and Medical Practice in Britain, 1865–1900* (Cambridge: Cambridge University Press, 2000).

The scholarly literature on animals has been distinct from that of veterinary history. Following Claude Lévi-Strauss' famous injunction that animals are "good to think with," historians have increasingly begun to use nonhuman creatures in their work. Few scholars have viewed animals as central historical actors, however, or deemed the relationships between humans and animals to be worthy of serious historical analysis. Rather, animals have usually provided case studies that illustrated authors' contentions about *human* society and culture. Animals have been portrayed as foils for what it means to be human, as representations of human cultural meaning, or as components of the background against which human dramas have unfolded. See, for example, Kathleen Kete's *The Beast in the Boudoir: Petkeeping in Nineteenth-Century Paris* (Berkeley: University of California Press, 1994) and Harriet Ritvo's *The Animal Estate: The English and Other Creatures in the Victorian Age* (Cambridge: Harvard University Press, 1987). Keith Thomas's *Man and the Natural World: A History of the Modern Sensibility* (New York: Pantheon Books, 1983) remains an important source for the history of attitudes toward animals. Historical controversies over animal experimentation and animal cruelty may be found in Richard French, *Antivivisection and Medical Science in Victorian Society* (Princeton, NJ: Princeton University Press, 1975); Susan E. Lederer, "The Controversy Over Animal Experimentation in America, 1880–1914," in Nicolaas A. Rupke, ed., *Vivisection in Historical Perspective* (London: Routledge, 1990); Keith Tester, *Animals and Society: The Humanity of Animal Rights* (London: Routledge, 1991); and James Turner, *Reckoning with the Beast: Animals, Pain, and Humanity in the Victorian Mind* (Baltimore, MD: Johns Hopkins University Press, 1980). Environmental historians have also touched on historical human-animal relationships; see, for example, William Cronon, *Nature's Metropolis: Chicago and the Great West* (New York: Norton, 1991).

Sociologists, zoologists, and anthropologists have increasingly focused on human attitudes toward animals and the parameters of human association with animals. Some of these contain historical discussions. See James Serpell, *In the Company of Animals: A Study of Human-Animal Relationships* (Oxford: Basil Blackwell, 1986); Stephen Kellert, *The Value of Life: Biological Diversity and Human Society* (Washington, DC: Island Press, 1996), and "American Attitudes Toward and Knowledge of Animals: An Update," *International Journal of Studies of Animal Problems* 1 (1980): 87–119; Elizabeth Atwood Lawrence, *Hunting the Wren: The Transformation of Bird to Symbol, a Study in Human-Animal Relationships* (Knoxville: University of Tennessee Press, 1997); Jennifer Wolch and Jody Emel, eds., *Animal Geographies: Place, Politics, and Identity in the Nature-Culture Borderlands* (London: Verso, 1998); and Chris Philo and Chris Wilbert, eds, *Animal Spaces, Beastly Places: New Geographies of Human-Animal Relations* (London: Routledge, 2000).

PRIMARY SOURCE MATERIAL FOR THE HISTORY OF VETERINARY MEDICINE comes in many forms. The sources most readily available to historians include journals (both scientific and popular publications), government documents, and archives (oral history is available for more recent events, of course).

The most enduring journal of veterinary medicine in the United States is the *Journal of the American Veterinary Medical Association (JAVMA)*. Prior to 1915, this journal was known as the *American Veterinary Review* (founded 1877). The editor for its first quarter-century was Alexandre Liautard, a French-trained veterinarian interested in promoting that nation's germ ideas and a strong federal veterinary public health service. Because *AVR/JAVMA* has occupied a primary place in circulation numbers and authority over the years, its editors have been influential veterinary leaders. The other major nineteenth-century journal (1880–1903) was the *Journal of Comparative Medicine and Veterinary Archives*. After the turn of the twentieth century, Cornell University began publishing its quarterly journal, *Cornell Veterinarian*. Joining it and the *AVR/JAVMA* were the very popular practitioner-oriented publication the *American Journal of Veterinary Medicine* and later J. V. Lacroix's *North American Veterinarian* (focusing, in accordance with its editor's interest, on small-animal medicine and surgery). The *American Journal of Veterinary Medicine* was renamed *Veterinary Medicine* in the 1920s, and it continued to provide practical information written by and for practitioners (mostly on livestock issues). *North American Veterinarian* was renamed *Modern Veterinary Practice* in the 1960s and its orientation shifted toward large-animal-practice issues. The growth of veterinary research in the 1930s stimulated the founding of a research journal, the *American Journal of Veterinary Research (AJVR)*, in 1940. This journal, and the growth of research in this time period, deserve more attention from historians. After World War II, the above journals were joined by many competitors and spinoff specialty publications, including the *Journal of the American Animal Hospital Association, Veterinary Economics, Veterinary Medicine–Small Animal Clinician*, and many more too numerous to list. Historians of the late twentieth century will also wish to consider Internet-based sources of information for veterinarians, starting with the American Veterinary Medical Association's *NOAH* site.

Veterinary medicine, more so than human medicine, has been closely associated with municipal, state, and federal agencies for much of its existence as a recognized profession. Therefore, government documents and archives are a fundamental source of information. The following list, while not comprehensive, suggests some usually fruitful areas in which to search for information. Municipal archives and legal records include reports of city boards of health; nuisance laws; ward censuses of animals; city directories; statements of properties; veterinary license registers; and health ordinances dealing with milk, meat, dog waste, rabies, and other issues. Useful state records include livestock board minutes, proposed legislation, and published informational pamphlets; reports from state research laboratories; reports from the office of state veterinarian; state agriculture department records and publications; and state laws pertaining to the movement of animals and public health concerns.

A cornerstone of research on American veterinary medicine is the archive of the federal Bureau of Animal Industry (BAI) (1884–1953). These records are held as Record Group 17 at the National Archives and Records Administration (NARA), College Park, Maryland. While this is a large and rich collection, it is also uncatalogued and frustrating to use. An inefficient numerical filing system used by the bureau and a rudimentary finding guide are the only organizational features. Scholars intending to use this archive should contact a NARA archivist and other scholars familiar with it, because they often trade information on the contents of the many boxes in this collection. Significant material relating to veterinary education, outbreaks of disease, meat inspection, regulation of animal populations, and scientific research will be found in this collection. The BAI issued annual reports in which both administrative details and the results of research were published. In addition, BAI *Bulletins* were printed periodically as a conduit for researchers to report their findings.

The U.S. Department of Agriculture (USDA) and the federal census are two other important sources of information. Although census statistics suffer from multiple sources of error, they demonstrate change in what officials believed was worth measuring, how they proposed to measure it, and crude numerical trends over time. The USDA published the *Yearbooks of Agriculture*, reports, and bulletins detailing research on a number of animal diseases and problems separately from the Bureau of Animal Industry documents. The National Agricultural Library in Beltsville, Maryland, holds many of these publications. The federally funded, state-run agricultural experiment stations were also major sources of research into animal health problems, and they employed veterinarians. Experiment station bulletins and related sources deserve more attention from historians.

University archives provide a rich and often untapped source of primary and early secondary materials, both manuscript and published. Land-grant colleges and other universities with veterinary schools (past or present) are most likely to hold materials. This is not universally true; for example, the University of Pennsylvania archives contain very little from the university's school of veterinary medicine despite its long tenure there. On the other hand, Cornell University holds large and significant collections in both the Kroch Library and Flower-Sprecher Veterinary Library. The former is the home of materials collected by deans of the veterinary college from the late nineteenth through the mid-twentieth centuries: the V. A. Moore papers (voluminous and a good place to start), some papers from James Law, the Pierre Augustus Fish papers, and the William A. Hagan papers. Other highlights of this library's collections include the papers of Fred Lucius Kilborne (co-discoverer of the vector transmission of Texas cattle fever), Raymond Russell Birch (practitioner), Simon Henry Gage (eminent anatomist and faculty member), and many more. Practitioner collections contain valuable documents such as day books, tuberculosis inspection records, and other important sources of information about daily practice. The Flower-Sprecher Library has a rare book room, the tabulated materials that Ellis Pierson Leonard used to write his histories of veterinary medicine at Cornell, and unprocessed archival materials from the early days of the institution.

Besides holding one of the strongest collections of agricultural and animal science journals in the nation, Iowa State University houses the papers of former Secretary of Agriculture James A. Wilson, researcher Marion Dorset, faculty members Edward Benbrook and Margaret Wragg Sloss, and many others. Iowa State's collection is particularly strong in the area of meat production and inspection, including the papers of the Livestock Conservation Institute and the Rath Packing Company, the Clarence H. Pals papers, and the Carl Laurel Telleen papers. Veterinary historian J. F. Smithcors has donated his extensive collection of books, papers, and other materials to Washington State University; this library also holds the archives of the Association of Women Veterinarians. While the aforementioned collections are the best places to begin, many other universities will hold materials of interest. Finally, the veterinary community is a close-knit one, with many veterinarians tracing the profession back through their families for generations. Private individuals are often interested in sharing their old books, manuscript collections, and instruments with historians. For the events of the past 30 years, historians are likely to be granted interviews by veterinarians, many of whom are quite interested in the history of their profession and their part in it.

Primary source materials on veterinarians' patients and the industries associated with them are far flung and best organized according to animal species. As this book has argued, understanding the social position of pet animals requires an examination of the popular literature, the work of animal protection organizations, and industries producing pet foods and other products. The major journals for fanciers include *Dog World* and *Cat Fancy;* in the earlier years of the twentieth century, *Country Gentleman* paid particular attention to dogs. Publications about pet animals have multiplied in the past 50 years. Animal protection organizations published journals including the *Journal of Zoophily / Starry Cross*, the *National Humane Review*, and *Our Dumb Animals;* early archival materials are sporadically available by contacting the organizations. Pet food companies consolidated themselves into the Pet Food Institute, which issued reports containing useful information such as estimates of pets owned and food purchased. The American Kennel Club and individual breed associations also have records of interest to historians. Veterinarians interested in dogs and cats read the *North American Veterinarian* in the early part of the century and such journals as *Veterinary Medicine–Small Animal Practice* later. The American Animal Hospital Association is a good source of information on animal hospitals, medical and surgical procedures, and the specialized concerns of small-animal practitioners. Pharmaceutical companies such as Lederle Laboratories also regularly published guides for veterinarians on small-animal medicine and surgery.

Horses may now be considered pets, and the equine fancying literature has become quite extensive over the past 30 years (including journals such as *Western Horseman, Practical Horseman,* the health-oriented *Equus,* and many more). Horses as working animals earlier in the century may be occasionally found in animal protection and welfare literature (including the popular reform journal *Charities* during the 1900s and 1910s). Business records and municipal records from this early time period often list horses, grooms, drivers, and supplies. An excellent early collection of practical literature devoted to working horses is the Fairman Rogers Collection, New Bolton Center Library, University of Pennsylvania (Kennett Square). The American Association of Equine Practitioners is a good source of information about veterinarians' specialized interest in horses following the heyday of working-horse practice.

The literature on food-producing animals (chickens, cattle, swine, sheep, and goats) is the most extensive. From the production point of view, early published journals of importance include *Breeder's Gazette, Wallace's Farmer, Hoard's Dairyman,* and publications of the Chicago Board of Trade. Since World War II, journals targeting farmers and producers of certain species, such as *Poultry Tribune,* have multiplied. Pharmaceutical and feed companies published their own trade journals, such as *Feed Age,* and some may also allow historians access to archives from this period. As previously mentioned, land-grant universities such as Iowa State University are the best places to find selections of producer-oriented literature. Historians have done more to explore the issues surrounding food from consumers' point of view, and there are some excellent monographs on this subject, including Harvey Levenstein's *Paradox of Plenty: A Social History of Eating in Modern America* (New York: Oxford University Press, 1993).

Producers' trade journals should not be neglected as a source of information on consumer tastes as well. Food-producing animals have long been an interest of physicians and public health officials. Here, too, a number of monographs contain chapters or sections on milk or meat regulations and can direct historians to sources. (A good example is Jacqueline H. Wolf's *Don't Kill Your Baby: Public Health and the Decline of Breastfeeding in the 19th and 20th Centuries* [Columbus: Ohio State University Press].) Veterinary journals primarily interested in food-producing animals included the *American Journal of Veterinary Medicine,* renamed *Veterinary Med-*

icine, and *Modern Veterinary Practice.* The American Association of Bovine Practitioners and similar organizations are also sources of information on veterinarians interested in these animals.

This book represents but one journey through these sources. Historians should know that although the materials described here are widely dispersed, they have been little used. It is my hope that this book encourages others to make use of the rich resources available for the study of comparative medicine and the history of human-animal relationships.

Index

Actors, historical, animals as, 4
Agricultural depression, 47, 60
Agricultural experiment stations, 61, 99
Agriculture, intensive animal, 92, 100, 113, 145
American Animal Hospital Association, 136
American Antivivisection Society, 124
American Association of Feline Practitioners, 136
American Humane Association, 141
American Public Health Association, 77, 88
American Veterinary Medical Association (AVMA), 7, 25, 55, 57, 106, 123, 127, 136
American Veterinary Review, 15, 28, 81
Anderson, Clinton P., 98
Anderson, W. A., 106
Animal experimentation, 126–8
Animal populations, 17, 46
Animal protection organizations, 36, 44, 122, 139, 150
Animal science, 99
Animal welfare, 36, 44, 73, 111–3, 116, 127–8, 137, 141, 145; state laws concerning, 36, 143
Anthropomorphization, 3
Antibiotic growth effect, 107–9

Antibiotics, 97–8. *See also* Antibiotic growth effect
Antivivisection movement, 124, 130, 150; federal legislation and, 125–7; gender and, 150–1; and veterinary medicine, 124–8, 149–51
Armour, J. Ogden, 86
Association of Veterinary Faculties, 57
Automobiles, 41, 52

BAI. *See* Bureau of Animal Industry
Barkan, Ilyse, 73
Bicycles, 40
Billings, F. S., 31
Blair, Sheryl, 26
Bovine somatotropin (BST), 152
Bovine spongiform encephalopathy (BSE), 152–3
Buckingham, David, 125–7
Bureau of Animal Industry (BAI; USDA), 7, 55, 93, 125, 127–8, 131; dissolution of, 7, 110; as employer of veterinarians, 56–8; establishment of, 12, 16, 29; and federal meat inspection, 82–8; regulation of veterinary education, 33, 55–8; Texas cattle fever research and, 30
Butler, Tait, 25

Carnation Company, 145
Cart-horses, 39

Cats, 132–3, 136; in cities, 22
Cattle: beef, 19; dairy, 19, 21, 66
Center for Science in the Public Interest, 151
Certified milk, 68
Chain, Ernst, 97
Cities, animals in, 18, 22. *See also* Urban reform
"City Practical," 24
Coccidiosis, 104, 106
Commercial Solvents (firm), 97
Communism, 92, 99
Companion animals, 8, 115, 146–7, 153–4; veterinary practice and, 8, 59, 114, 123, 129
Comparative medicine, definition, 4
Cornell University, 29
County extension agents, 96, 99
Crane, Caroline Bartlett, 74–7, 84, 88, 143
Crawford, J. Elliott, 129
Cronon, William, 146
Curtice, Cooper, 30

Depression (1930s), 93–5, 129, 135
Dickens, Charles, 19
Dinsmore, R. J., 11
Disease, definition of, 88–9, 111
Distemper, 132
Dogs, 115, 130–2; and Alaska diphtheria epidemic, 117; in cities, 22; heroic narratives about, 117; in motion pictures, 118
Domestic animal economy, 61; definition, 2–3
Dykstra, Ralph R., 95
Dyson, O. E., 77

Embalmed beef scandal, 84
Environmental Defense (advocacy group), 152
Environmental history, 4
Europe, veterinary education in, 28, 78

"Factory farms," 8, 110–1, 152. *See also* Agriculture, intensive animal

Feist, A. A., 123, 129, 139
Fire horses, 42
Fish, Pierre A., 59, 129
Fleming, Alexander, 97
Flexner Report, 57
Flynn, J. C., 122, 136
Food-producing animals, 144–6, 151–3; owners and producers of, 64, 69; veterinary practice and, 13, 61. *See also* Cattle; Goats; Poultry; Sheep; Swine
Foot-and-mouth disease, 32
Frink, W. E., 135

Gay, Carl, 41
General Federation of Women's Clubs, 76
Germ-free animals, 109
Germ theories, 28, 31
Glanders, 24, 80
Goats, 19
Goldhaft, Arthur, 26, 27, 33, 50, 93, 103, 105
Goler, George W., 63
Great Epizootic (1872), 24
Greatest Trust in the World, The (Russell), 85

Hagan, William A., 97, 98
Hall, Maurice C., 125–8, 149–51
Herd health, 61, 92, 111
"Heroic" treatments, 14
Herriot, James (Alfred Wight), 48
Hog cholera, 24; federal eradication of, 93, 95
Hookworm, 126
"Horse doctors," 12
Horse to motor power transition, 8, 37, 40–7
Horses, 37, 147–8; American breeds of, 39; in cities, 23, 26; class hierarchy of, 38–40; fire, 42; importance to veterinary practitioners, 15, 37, 47; as symbol of antimodernism, 41
Hoskins, W. Horace, 74

Hospitals, animal, 59, 134–6
Houck, Ulysses G., 32
Hoy, Suellen, 73
Hughes, D. Arthur, 86
Humane societies. *See* Animal protection organizations
Humanitarianism, 46, 128, 142–4, 146–8; pets as symbols of, 2, 146–7

Industries: animals as raw materials in, 18, 142; workers in animal, 21, 142
Irwin, Helen Richt, 140

Jaques, William K., 86
Jennings, Robert, 12
Jen-Sal, 132
Jones, C. Hampson, 72
Journal of the American Veterinary Medical Association, 98, 123, 132, 135
Jukes, Thomas, 108
Jungle, The (Sinclair), 84–5

Keen, W. W., 149
Ken-L Ration, 119
Kiernan, John, 94
Kilborne, Fred L., 30
Kitchen, Herminie, 98
Klein, Louis, 71
Koch, Robert, 28, 31, 78
Koen, J. S., 96

Laboratory animals, 124
Laboratory sciences in veterinary medicine, 54
Lassie, 118, 120
Law, James, 29
Lederle Laboratories, 97, 104, 107, 132, 133
Legal status of animals, 5, 153–4
Liautard, Alexandre, 16, 28, 96
Lister, Joseph, 28
Little, George Watson, 132
Livestock. *See* Food-producing animals
Lowe, William H., 28, 77

"Mad cow disease." *See* Bovine spongiform encephalopathy
Mallein, 32
Marx, Karl, 1
Massachusetts Society for the Prevention of Cruelty to Animals, 143
Massachusetts Veterinary Medical Association, 131
Meat: export of, 29; inspection, 74, 82, 87, 106, 151; and tuberculosis, 88
Meat inspection act of 1891, 83
Meat inspection act of 1906, 57, 87, 112
Meat packers, Chicago, 76, 83
Melvin, Alonzo D., 56, 85
Merck, Sharp and Dohme, 97, 104, 105
Merillat, Louis A., 95
Metropolitan Life Insurance Company, 145
Milk: certified, 68; and children, 63, 72, 152; municipal regulations concerning, 65–72; and tuberculosis, 68. *See also* Pasteurization
Modernity, 6, 142, 149; as moral problem, 41, 144
Modern Veterinary Practice, 110
Mohler, John R., 56, 149–51
Moore, Veranus A., 49, 54, 56, 143
Morris, Mark, 123, 125, 132, 136

National Board of Health, 16
National Cattlemen's Beef Association, 152
National Dog Food Manufacturer's Association, 119
National Grange, 71
Newcastle disease, 106
Nietzsche, Friedrich, 1
Nuisances, animal-related, and public health concerns, 21, 23
Nutritional science, 102

Osteen, O. L., 106

"Packingtown" (Chicago), 18, 21
Paine, Thomas, 142

Parke-Davis Company, 97, 131
Parker, F. F., 123
Pasteur, Louis, 28
Pasteurization, 68, 89
Pearson, Leonard, 33, 36, 74, 78, 85, 94
Pellagra ("black tongue"), 125
"Pennsylvania plan" for tuberculosis
    eradication, 79
Pennsylvania Veterinary Medical Associ-
    ation, 78, 81
Pets. *See* Companion animals
Pharmaceutical industry, 101, 104
Physicians, 71, 73, 126, 143, 150
Pitman-Moore, 132, 133
Pleuropneumonia, bovine, 29, 80
Poultry, 101
Poultry Tribune, 98, 105
Preventive medicine, 61, 106, 109, 133
Prices, animal, 26, 46, 60
Price-value theory, 5
Progressive Era, 143
Public health, 95; women as reformers
    of, 64, 72, 143
Pullorum disease, 103

Rabies, 23, 130–2
Rin-Tin-Tin, 118, 120
Ritvo, Harriet, 146, 149
Rockefeller Institute, 133
Roosevelt, Eleanor, 95
Roosevelt, Theodore, 84
Rossiter, Margaret, 100
Rowley, Francis, 143
Rural areas, animals in, 17, 43, 61
Rush, Benjamin, 143
Russell, Charles Edward, 84

Salmon, Daniel E., 16, 29, 56, 82, 85
Schalk, A. F., 26
Schilling, S. J., 60
Schmoker, E. A., 107
Schoening, Harry W., 131
Schofield, F. W., 98
Scholtmeijer, Marian, 142

Schreck, Horst, 135
Sheep, 19
Sherman Anti-Trust Act, 84
Simmel, Georg, 1, 5
Sinclair, Upton, 84
Slaughterhouses, 74
Smith, Adolphe, 84
Smith, Theobald, 30, 89
Smithcors, J. F., 112
Spratt's Biscuit Company, 119
Staley, R. M., 49
Stalheim, O. H. V., 96
Stein, Barbara, 136
Stockyards, 18
Straus, Nathan, 67
Streeter, A. H., 27
Sulfonamides, 97, 104
Swabe, Joanna, 112
Swine, 19, 109–10

Taylor, Alonzo, 78, 82
Teigen, Philip M., 18
Terhune, Albert Payson, 116
Texas cattle fever, 30
Thompson, E. P., 3
Tractors, 43
Tuan, Yi-Fu, 146
Tuberculin, 32, 69, 70, 79, 81, 93–4, 130
Tuberculosis, 24, 79; federal eradication
    of, in cattle, 71, 89, 93–4; and meat, 88;
    milk regulations and, 68, 69
Twentieth century, 142; human-animal
    relationships in, 3

Udall, D. H., 70
United States Department of Agriculture
    (USDA), 1. *See also* Bureau of Animal
    Industry
United States Department of the Trea-
    sury, 29
United States Supreme Court, 70
Urban reform, 22, 24, 73
Urban-rural boundaries, 65, 71

Vaccines, 130–3; for distemper, 132; for rabies, 130–2

Value: definition of, 1; and money, 5; sentiment as basis for, 124, 137–9; veterinarians as judges of, in animals, 6, 158. *See also* Price-value theory

Veterinarians: as mediators of human-animal relationships, 1, 6; as practitioners, 15, 25, 59, 80, 90, 93–5, 104, 107, 110, 115–6, 132, 140; as scientists and researchers, 28, 104, 106, 130

Veterinary Medicine, 53, 107, 135

Veterinary profession, 147–51; in cities, 25; culture of, 11, 124, 139, 151; and dependence on value of animals, 51; development of, 8; education and, 11, 15, 26, 33, 48, 55–8; and federal government, 57, 94, 99; gender and, 12, 150–1; in the 19th century, 11; and public health issues, 61; state practice acts and, 15, 48

Veterinary schools: associated with universities, 49, 55; curricula of, 26, 48, 54–5, 61; proprietary or private, 49, 55

Victorian culture, and animals, 116

Vivisection Investigation League, 125–8

Waksman, Selman, 98

Weber, Max, 6

West, Elliott, 4

Wheeler, A. S., 74, 77

White, George R., 77

Wilson, James A., secretary of agriculture, 57, 85

Wilson, Woodrow, 41

Wolf, Jacqueline, 145

Women: as animal owners, 139; as veterinarians, 13, 139–40. *See also* Public health: women as reformers of; Veterinary profession: gender and

Women's Health Protective Association, 74

Work horse parades, 35–7

World War I, 46–7, 117

World War II, 98

Worster, Donald, 3

Young, George A., 109

Zelizer, Viviana, 5